Patent Aggregating Companies

Frauke Rüther

Patent Aggregating Companies

Their strategies, activities and options for producing companies

Frauke Rüther
St. Gallen, Switzerland

Doctoral thesis, University of St.Gallen, 2012

ISBN 978-3-8349-4454-2   ISBN 978-3-8349-4455-9 (eBook)
DOI 10.1007/978-3-8349-4455-9

The Deutsche Nationalbibliothek lists this publication in the Deutsche Nationalbibliografie; detailed bibliographic data are available in the Internet at http://dnb.d-nb.de.

Springer Gabler
© Springer Fachmedien Wiesbaden 2013
This work is subject to copyright. All rights are reserved by the Publisher, whether the whole or part of the material is concerned, specifically the rights of translation, reprinting, reuse of illustrations, recitation, broadcasting, reproduction on microfilms or in any other physical way, and transmission or information storage and retrieval, electronic adaptation, computer software, or by similar or dissimilar methodology now known or hereafter developed. Exempted from this legal reservation are brief excerpts in connection with reviews or scholarly analysis or material supplied specifically for the purpose of being entered and executed on a computer system, for exclusive use by the purchaser of the work. Duplication of this publication or parts thereof is permitted only under the provisions of the Copyright Law of the Publisher's location, in its current version, and permission for use must always be obtained from Springer. Permissions for use may be obtained through RightsLink at the Copyright Clearance Center. Violations are liable to prosecution under the respective Copyright Law.
The use of general descriptive names, registered names, trademarks, service marks, etc. in this publication does not imply, even in the absence of a specific statement, that such names are exempt from the relevant protective laws and regulations and therefore free for general use.
While the advice and information in this book are believed to be true and accurate at the date of publication, neither the authors nor the editors nor the publisher can accept any legal responsibility for any errors or omissions that may be made. The publisher makes no warranty, express or implied, with respect to the material contained herein.

Printed on acid-free paper

Springer Gabler is a brand of Springer DE. Springer DE is part of Springer Science+Business Media.
www.springer-gabler.de

Preface

This thesis is the result of my research carried out at the Institute of Technology Management at the University of St. Gallen. Particular thanks and gratitude go to my supervisor, Professor Oliver Gassmann. Oliver, I thank you that you believed in me, often more than I believed in myself. And thank you for giving me the opportunity and the freedom to pursue my academic interests and aspirations. I would also like to thank Professor Beat Bernet for co-supervising my thesis.

I would also like to thank Professor Beth Webster for supporting my research sabbatical at the Intellectual Property Research Institute of Australia, University of Melbourne. Beth, thanks for always pushing and encouraging me. Even though I often felt like a deer caught in the headlights, your constant support made a real difference to my work. I am deeply grateful to the Swiss National Science Foundation for providing me with financial support during my time in Melbourne.

For contributing valuable input to this work, I am thankful to several colleagues and students at the Institute of Technology Management. Thank you Martin Bader, Matthias Bitzer, Lukas Burkhardt, Pascal Oesch, Carsten Vollmar, Bastian Widenmayer, Marco Zeschky, and Nicole Ziegler who was not only a challenging discussion partner but also read and commented on the entire manuscript. Also thanks to all my other team members for the great time we spent together, especially the non-academic hours in academic environment – long live the Bergfest on Spritwoch and its inventors Sascha Friesike and Michael Daiber. I would also like to thank Ursula Elsässer for her constant support and ever-present sympathetic ear.

Without my friends, I would not have finished my research, and I am deeply grateful that they are still my friends despite my moods, temper, and sometimes long spans of silence. Rebecca Fröhlich – Hasisister and Schager of my life, thank you that you are my best friend… always… anytime… anywhere. I just say 17 and R.F.F.R. Sonja Höh – Zuckerpuppe, I never thought I could survive without your culinary support, but the moral support you gave me the last years rescued me from more severe things than an empty stomach. Bettina Maisch – my Silvester-Mate, my favorite career change will always be 'Klofrau', because I met an amazing woman when I performed the job of proctoring an exam. Thank you for everything you did and said to me, especially on New Year's Eve 2010. Sabine Ravimonica – Puschel, words are not enough to show you my gratitude for your friendship and your Genihase-sitting; you were my sunshine in St. Gallen's rain. Nicole Ziegler – LOML, you are now the impossible high standard for all my future colleagues. I will miss you for the rest of my career. Thanks for being there, listing, and going with me through all ups and downs of an ITEM life. I also raise my glass to the Wednesday all-stars Julia Bendul, Jochen Binder, Antonia Erz (– Tonlein, thanks so much for your effort that Lilifee could come home and all the time you spent with me), Dennis Herhausen, Philip Schnaith (– Don Philipo, http://www.youtube.com/watch?v=H1BKxYyJJJ4 for all the http://www.youtube.com/watch?v=smh834dsYu8 and http://www.youtube.com/watch?v=KklLsdWaoBU), and Friederike Wolter, and say thank you for all the memorable hours and that nobody forced me to cook.

A tower of strength has always been my aunt Dr. Magrit Ritterhoff. Tante Magrit, thank you for everything. You cannot imagine what your steady presence means to me. I am especially thankful to my brother Cord Rüther, his wife Birthe and my beloved niece and nephews Espe, Lasse, and Tjorven. I am so grateful that you never questioned me and I always found refuge with you during my time in St. Gallen. Last but not least, I would like to thank the most important persons in my life, my parents Almut and Hans-Heinrich Rüther. Mama und Papa, auch wenn wir räumlich weit entfernt waren hat Eure Liebe, Zuversicht und immerwährende Unterstützung dafür gesorgt, dass Ihr mir in meinem Herzen ganz nah wart. Ihr fangt mich auf wenn ich zu fallen drohe und ihr habt mir Wurzeln und Flügel gegeben. I dedicate this thesis to my family.

Melbourne and St. Gallen, April 2012   Frauke Rüther

Index

Contents IX
Figures XIII
Tables XV
Abbreviations XVI
Abstract XVII
Zusammenfassung XVIII

1   Introduction 1
    1.1   Motivation 1
    1.2   Research objectives and questions 12
    1.3   Terms and definitions 13
    1.4   Research concept and methodology 15
    1.5   Thesis structure 21

2   Leveraging companies' patent portfolios – State of the art 24
    2.1   Fundamentals of patent management 24
    2.2   Options to leverage patent portfolios 28
    2.3   Third parties as enablers of transactions 37
    2.4   Reference framework 45

3   Exploring the phenomenon of patent aggregating companies 48
    3.1   Setting of patent aggregating companies 48
    3.2   Process of patent aggregation 54
    3.3   Strategies of patent aggregating companies 63
    3.4   Summary 68

4   Potentials offered by patent aggregating companies 69
    4.1   External potentials offered by patent aggregating companies 70
    4.2   Internal potentials offered by patent aggregating companies 80
    4.3   Summary 91

5   Typology of patent aggregating companies 93

5.1	Four archetypes of patent aggregating companies	94
5.2	Archetype 1 – Merchant	99
5.3	Archetype 2 – Gardener	115
5.4	Archetype 3 – Collector	127
5.5	Archetype 4 – Patron	140
5.6	Summary and evaluation of potentials	151
6	Leveraging patent portfolios by utilizing patent aggregating companies	154
6.1	Managing the utilization of patent aggregating companies	154
6.2	Development of patent aggregating companies	166
6.3	Meeting demand for learning effect as driving factor	171
6.4	Summary	179
7	Conclusion	181
7.1	Contribution to management theory	181
7.2	Implications for management practice	185
7.3	Further research and trends	187
References		192
Appendix		211

Contents

Contents IX
Figures XIII
Tables XV
Abbreviations XVI
Abstract XVII
Zusammenfassung XVIII

1 Introduction 1
   1.1 Motivation 1
      1.1.1 The market for technologies and the emergence of a new player 1
      1.1.2 Practical challenges in leveraging corporate patent portfolios 5
      1.1.3 Deficits in current research 8
   1.2 Research objectives and questions 12
   1.3 Terms and definitions 13
   1.4 Research concept and methodology 15
   1.5 Thesis structure 21

2 Leveraging companies' patent portfolios – State of the art 24
   2.1 Fundamentals of patent management 24
      2.1.1 Reasons why firms patent 25
      2.1.2 Reasons why companies buy patents 27
   2.2 Options to leverage patent portfolios 28
      2.2.1 Internal and external exploitation of patents 28
      2.2.2 Impediments to optimally leverage patent portfolios 34
   2.3 Third parties as enablers of transactions 37
      2.3.1 Bridging patent supply and patent demand 38
      2.3.2 Non-practicing entities and their intermediation of patent transactions 41
   2.4 Reference framework 45

3 Exploring the phenomenon of patent aggregating companies 48
   3.1 Setting of patent aggregating companies 48
      3.1.1 General information 48

    3.1.2   Venture creation and funding of patent aggregating companies 51

  3.2   Process of patent aggregation 54

    3.2.1   Selection of patents 55
    3.2.2   Structuring of patent portfolios 57
    3.2.3   Additional value adding activities 59
    3.2.4   Exploitation of patents 60

  3.3   Strategies of patent aggregating companies 63

    3.3.1   Basic strategy I: Generate revenues 63
    3.3.2   Basic strategy II: Serve an objective 64
    3.3.3   Eight business models of patent aggregating companies 66

  3.4   Summary 68

4   Potentials offered by patent aggregating companies 69

  4.1   External potentials offered by patent aggregating companies 70

    4.1.1   Potentials for risks reduction 70
    4.1.2   Potentials for market fostering 74
    4.1.3   Potentials for resource enhancement 77

  4.2   Internal potentials offered by patent aggregating companies 80

    4.2.1   Potential for market interaction 80
    4.2.2   Potentials for cost effectiveness 83
    4.2.3   Potentials for decision making 89

  4.3   Summary 91

5   Typology of patent aggregating companies 93

  5.1   Four archetypes of patent aggregating companies 94

  5.2   Archetype 1 – Merchant 99

    5.2.1   Patent trading fund's characteristics 99
    5.2.2   Patent trading fund's case study: Alpha Patentfonds 101
    5.2.3   Patent acquisition company's characteristics 106
    5.2.4   Patent acquisition company's case study: Intellectual Ventures 109

  5.3   Archetype 2 – Gardener 115

    5.3.1   Royalty monetization company's characteristics 115
    5.3.2   Royalty monetization company's case study: Pete Invest MedTech. 118
    5.3.3   Patent incubating fund's characteristics 121
    5.3.4   Patent incubating fund's case study: Patent Select 123

5.4 Archetype 3 – Collector 127

- 5.4.1 Patent enforcement company's characteristics 127
- 5.4.2 Patent enforcement company's case study: Acacia Research 130
- 5.4.3 Defensive patent aggregator's characteristics 134
- 5.4.4 Defensive patent aggregator's case study: Allied Security Trust 137

5.5 Archetype 4 – Patron 140

- 5.5.1 Patent pooling company's characteristics 140
- 5.5.2 Patent pooling company's case study: MPEG LA 142
- 5.5.3 Non-commercial patent aggregator's characteristics 145
- 5.5.4 Non-commercial patent aggregator's case study: Golden Rice PDP 147

5.6 Summary and evaluation of potentials 151

6 Leveraging patent portfolios by utilizing patent aggregating companies 154

6.1 Managing the utilization of patent aggregating companies 154

- 6.1.1 Value generating options and patent aggregating companies 154
- 6.1.2 Constraints in utilizing patent aggregating companies 157
- 6.1.3 Framework for the utilization of patent aggregating companies 163

6.2 Development of patent aggregating companies 166

- 6.2.1 Trend 1: From aggregation of interest to aggregation of investments 167
- 6.2.2 Trend 2: Responses to organized patent enforcement 169
- 6.2.3 Trend 3: From enforcement agents to innovation intermediaries 170

6.3 Meeting demand for learning effect as driving factor 171

- 6.3.1 Monetary benefits of utilizing patent aggregating companies 171
- 6.3.2 Non-monetary benefits of utilizing patent aggregating companies ... 173
- 6.3.3 Benefits depend on the type of patent aggregating company 174

6.4 Summary 179

7 Conclusion 181

7.1 Contribution to management theory 181
7.2 Implications for management practice 185
7.3 Further research and trends 187

References 192
Appendix 211

Figures

Figure 1: Relevant literature streams and research gap 9
Figure 2: Research questions 12
Figure 3: Players and relationships in the patent aggregating ecosystem 14
Figure 4: Structure of the thesis 23
Figure 5: Map of value generating options for leveraging patent portfolios 34
Figure 6: Transfer of patents and technology market intermediaries 39
Figure 7: Reference framework to analyze patent aggregating companies 46
Figure 8: Year of formation and geographic location of sample companies 49
Figure 9: The different paths of venture creations 53
Figure 10: The process of patent aggregation 55
Figure 11: Patent exploitation options of patent aggregating companies 61
Figure 12: Business models of patent aggregating companies and their strategies 67
Figure 13: Overview of external and internal potentials 69
Figure 14: External potentials for risks reduction 70
Figure 15: External risks of R&D 71
Figure 16: External potentials for market fostering 75
Figure 17: External potentials for resource enhancement 77
Figure 18: Internal potentials for market interaction 81
Figure 19: Internal potentials for cost effectiveness 84
Figure 20: Principal-agent problem and patent aggregating company approach 86
Figure 21: Screening and selection process of Patent Select 87
Figure 22: Internal potentials for decision making 89
Figure 23: Summary of patent aggregating companies' potentials 92
Figure 24: Typology of patent aggregating companies 97
Figure 25: Summary of patent trading funds 101
Figure 26: Structure of the organization and relations of participants 103
Figure 27: Selection of patents in the structuring phase of the patent aggregating
 process of Alpha Patentfonds II 105
Figure 28: Summary of patent acquisition companies 108
Figure 29: Process of IP to EPS^{TM} transaction 112
Figure 30: Summary of royalty monetization companies 117
Figure 31: Summary of patent incubating funds 123

Figure 32: Exemplary structure of the organization and relations of participants illustrated on the investment fund Patent Portfolio I.................................. 125

Figure 33: Summary of patent enforcement companies 130

Figure 34: Sales and EBIT of Acacia from 2003 to 2010, in million USD................ 131

Figure 35: Summary of defensive patent aggregators 136

Figure 36: Structuring phase of Allied Security Trust.. 139

Figure 37: Summary of patent pooling companies... 142

Figure 38: Summary of non-commercial patent aggregators 147

Figure 39: Patent aggregating process of Golden Rice PDP 150

Figure 40: Value generating options that can include patent aggregating companies 155

Figure 41: Constraints that affect the utilization of patent aggregating companies ... 158

Figure 42: Management framework for utilizing patent aggregating companies 164

Figure 43: The three major trends that drive the evolution of patent aggregating companies 167

Figure 44: Resulting benefits for the original patent owner 175

Tables

Table 1: Empirical base of research 19
Table 2: Research sample of patent aggregating companies 20
Table 3: Evaluation of potentials by business model 153

Abbreviations

EPO	European Patent Office
EU	European Union
EUR	Euro
FDA	US Food and Drug Administration
FTO	Freedom to Operate
IP	Intellectual Property
IPR	Intellectual Property Rights
IPO	Initial Public Offering
LLC	Limited Liability Company
NPE	Non-practicing Entity
M&A	Mergers and Acquisitions
MNE	Multinational Enterprise
n/a	Not Available
OUH	Only Used Here
PA	Patent Aggregating Activities
PAC	Patent Aggregating Company
R&D	Research and Development
RFID	Radio Frequency Identification Domain
SME	Small and Medium-sized Enterprise
SPV	Special Purpose Vehicle
US	United States
USD	United States Dollar

Abstract

Entering the post-industrial age, knowledge has become an important asset for sustained competitive advantage. Therefore, patents, which in their historical meaning protect technical knowledge, have moved from a legal matter to a strategic issue. They are now longer only used to protect companies' products and processes but have developed to a currency that facilitates the trade of innovation. Producing companies have recognized this shift and increasingly license or sell patents, often with only moderate success due to the lack of internal capabilities and impediments to the market for patents and technologies. In recent years, a new acquirer type has emerged. Patent buyers and licensees are no longer solely producing companies but also third parties that seem to have none of the traditional acquisition motives. Even though these third parties do not produce goods and therefore, do not need patents in their historical meaning, they acquire patents and aggregate patent portfolios. Until now, little is known about patent aggregating companies. Their strategies, activities, and their evolution over time, as well as how producing companies can utilize them to leverage their patent portfolios are the subjects of this thesis.

Due to scarce empirical insights into patent aggregating companies, this thesis applies a qualitative, case-study based research approach. Based on data on 27 patent aggregating companies, existing literature on patent management, the market for technology, and technology market intermediaries are extended by examining the strategies, activities, and business models of patent aggregating companies. The case study analysis reveals that patent aggregating companies have eight different motives to aggregate patents. Further, the analysis shows that patent aggregating companies differ significantly regarding the competencies and rewards they offer to the original patent owners. These differences allow for deriving four archetypes. In addition, the archetypes allow patent managers of producing companies that wish to optimize their patent leveraging activities to select a suitable patent aggregating company.

The results conceptualize patent aggregating companies for the first time and go beyond the general picture of patent aggregating companies as enforcement agents. Findings show that since the founding of the first patent aggregating company, the business models have changed and now fulfill the function of innovation intermediaries. The results offer significant managerial implications for the leveraging activities of patent portfolios.

Zusammenfassung

Historisch gesehen sind Patente juristische Titel, die das technische Wissen von Unternehmen schützen. Durch den ökonomischen Wandel und dem damit verbundenen Eintritt in die Wissensgesellschaft hat sich die Bedeutung von Patenten zunehmend verändert. Heute dienen sie Firmen verstärkt als Währung im Handel von Innovationen und Wissen. Trotz steigender Anzahl von Patenttransaktionen und Lizenzgeschäften haben viele Firmen allerdings immer noch Schwierigkeiten, diese erfolgreich durchzuführen. Gleichzeitig ist zu beobachten, dass produzierende Firmen nicht länger die einzigen Teilnehmer auf dem Patentmarkt sind, sondern dass auch Firmen ohne eigene Forschung, Entwicklung oder Produktion immer häufiger als Käufer von Patenten in Erscheinung treten. Obwohl Letztere scheinbar keine Patente benötigen, aggregieren sie grosse Patentportfolios. Sie werden daher als Patent Aggregatoren bezeichnet. Welche Strategien diese Patent Aggregatoren dabei verfolgen, welche Aktivitäten sie betreiben und wie sie entstanden sind ist bisher kaum untersucht. Auch wie produzierende Firmen Patent Aggregatoren für die effizientere Nutzung des eigenen Patentportfolios einsetzten können, ist unklar. Diese Punkte sind Gegenstand der Untersuchung in dieser Arbeit.

Aufgrund der wenigen Erkenntnisse zu Patent Aggregatoren wird in dieser Arbeit ein qualitativer, Fallstudien-basierter Forschungsansatz verwendet. Durch die Untersuchung von Strategien, Aktivitäten und Geschäftsmodellen von 27 Patent Aggregatoren wird die bestehende Literatur zu Patentmanagement, Technologiemarkt-Intermediären und dem Patentmarkt ergänzt. Die Fallstudienanalyse zeigt, dass Patent Aggregatoren acht unterschiedliche Motive beim Kauf von Patenten verfolgen. Zudem unterscheiden sie sich hinsichtlich ihrer Kompetenzen und der Entlohnung an den Patentbesitzer. Basierend auf diesen unterschiedlichen Ausprägungen konnten vier Archetypen von Patent Aggregatoren identifiziert werden. Diese Archetypen eignen sich auch als Hilfestellung für produzierende Unternehmen, um die Suche nach geeigneten Partnern für die eigene externe Patentverwertung zu vereinfachen.

In der vorliegenden Forschungsarbeit wird erstmals eine Konzeptualisierung der verschiedenen Ausprägungen von Patent Aggregatoren vorgestellt. Die Ergebnisse dieser Arbeit relativieren zudem die übliche Annahme, dass Patent Aggregatoren nur als Auftragskläger wirken und zeigen, dass sich diese Firmen von Auftragsklägern zu Innovationsintermediären entwickelt haben.

1 Introduction

Patent aggregating companies, that is, companies that do not produce physical goods but amass large patent portfolios, have emerged recently in the international market for patents and technologies. Until now, little has been known about their strategies, motives, and of how they have evolved over time. Producing companies are not yet aware of how they could interact with or react to patent aggregating companies. The following chapter introduces the phenomenon of patent aggregating companies and derives a definition for this type of company. Laying out practical challenges in corporate patent management helps to derive the research objective and the research questions. In addition, the research concept and the empirical sample are described.

1.1 Motivation

Entering the post-industrial age, knowledge has become an important asset for sustained competitive advantage (Barney, 1991; Grant, 1996; Kogut & Zander, 1993; Nonaka, Toyama, & Nagata, 2000; Nonaka, 1994). Knowledge can be distinguished between tacit (informal, unstructured, uncodified) and explicit (formal, structured, codified) (Polanyi, 1962, Polanyi, 1967). As knowledge becomes more explicit, intellectual property rights (IPR) can be applied to protect it. Therefore, IPR are explicit knowledge resources and the most visible type of knowledge (Nonaka et al., 2000). The most important high technology IPR are patents (Pitkethly, 2001). The following parts describe the relevance of patents for today's companies. The first part describes patents as a good transacted in the market for patents and technologies and the two different types of buyers interested in them: producing companies and companies that do not have production and research and development (R&D). The second part describes the challenges producing companies face in their patent management. The third part reveals the deficits in current research.

1.1.1 The market for technologies and the emergence of a new player

Patents are legal rights with a possible economic value. The patent system was created to give the owner of an invention the right to exclude third parties to sell or use the invention (EPO, 2009). On the one hand, this exclusion of others helps the inventor to monopolize rewards from R&D. On the other hand, society benefits because the patent discloses information that promotes the state of the art (Gassmann & Bader, 2011). Traditionally, producing companies have conducted R&D internally and set up closed innovation processes. However, during the last decades the environment companies operate in has dramatically changed. Shorter product and technology life cycles (Chesbrough, 2003b; Christensen, Olesen, & Kjær, 2005; Granstrand, 2004; Grindley & Teece, 1997); a growing awareness of knowledge (Harris, 2001; Nonaka et al., 2000; OECD Publishing, 1996); increased costs of R&D (Keupp & Gassmann, 2009; Reepmeyer, Gassmann, & Rüther, 2011); and global competition (Gassmann, 2006) have forced firms to change their innovation process and shift to more open models of innovation (Chesbrough, 2003b; Chesbrough, 2006; Gassmann, 2006). In this new era of open innovation patents, are no longer used only internally but serve as legal instruments to trade technologies (Arora & Gambardella, 2010b; Gambardella, Giuri, & Luzzi, 2007; Gans & Stern, 2003; Scotchmer, 2006). Firms increasingly license or sell patents to external partners (Anderson, 1979; Chesbrough, 2003a; Lichtenthaler, 2005, Lichtenthaler, 2007c; Parr & Sullivan, 1996), and a market for patents and technologies has emerged (Arora, Fosfuri, & Gambardella, 2001a; Guilhon, 2001; Lamoreaux & Sokoloff, 2007; Teece, 1981).

The market for technology is a broad term and denotes trade in technology disembodied from physical goods (Arora & Gambardella, 2010b). Two different forms of patent transactions are possible. On the one hand, patents can be licensed or sold in combination with the technology and knowledge of the firm. In this case, the term *external technology exploitation* or *technology transfer* is used as well (Anderson, 1979; Ford & Ryan, 1977; Marcy, 1979). On the other hand, the sole legal right of exclusion is transferred without any knowledge or other intellectual assets of the firm (Lichtenthaler, 2007c; McDonough III, 2006; Shrestha, 2010).

Even though markets for technology already existed at the beginning of the 20th century (Lamoreaux & Sokoloff, 2007), structured activities and growth have started to emerge in the last decades (Arora et al., 2001a; Guilhon, 2001). On behalf of the OECD, Sheehan, Martinez, and Guellec (2004) surveyed 105 firms in Europe (68 firms), North America (20), and Asia-Pacific (17, mostly from Japan). The interviewees state that in almost 60% of the analyzed companies, in- and out-licensing notably increased during the 1990s. Some pioneering companies achieve significant strategic and monetary benefits by trading or licensing patents (Rivette & Kline,

2000). A successful practice firm is *IBM Corp.* Through adopting an active licensing program, *IBM's* licensing revenues increased from a mere USD 30 million in 1990 to more than USD 1.2 billion in 2004 (Lichtenthaler, 2007b). In the 1980s, *Texas Instruments* changed its business strategy and focused on exploiting the portfolio of patents that it had accumulated. Many companies used the patents without permission. Therefore, *Texas Instruments* started to enforce the patents covering the basic design of integrated circuits, and generated large royalty revenues with this strategy. For instance, a licensing agreement with several Japanese companies netted *Texas Instruments* USD 1.5 billion in licensing revenue by 1993 (Kline, 2003). Another example for exploiting its patent portfolio successfully is *Dow Chemicals*. In 1993, *Dow* introduced a new corporate strategic roadmap that implied management to save costs and to leverage the patent portfolio more effectively. Results of this new strategy were savings in taxes and maintenance fees of USD 50 million and an increase in licensing revenues from USD 25 million to more than USD 125 million (Davis & Harrison, 2001).

Patent licensing activities have increased not only in single companies, but also on an overall basis. Athreye and Cantwell (2007) analyze the trend in worldwide royalty and licensing revenues between 1950 and 2003 and find that they rose dramatically in the late 1980s and through the 1990s. The authors estimated that the international royalty and licensing revenues increased from ca. USD 35 billion in 1990 to ca. USD 70 billion in 2000. Kamiyama, Sheehan, and Martinez (2006) found similar results. They analyze OECD data on international receipts IP (including patents, copyrights, trademarks) and find that the total payments increased from USD 8.3 billion in 1985 to USD 120 billion in 2004. More than 90% of all receipts went to the three major OECD regions: the European Union (EU), Japan, and the United States (US).

In addition to patent licensing activities, patent sales activities have also increased. Due to their private nature, these transactions are hard to quantify (Arora & Gambardella, 2010a; Monk, 2009). Therefore, reliable data on the size of patent sales are not available but professionals agree that patent sales have become more common and are steadily increasing (e.g., Aronoff, 2011; Laurie, 2007; Pluvinage, 2011; Wild, 2010a). Using the USPTO Patent Assignment Database, Serrano (2010) shows that 13.5% of all granted patents are traded at least once over their life cycle. The study shows that patents covering technologies in the mechanical field are transferred far less than patents covering technologies in the field of drugs and medical.

For several reasons, producing companies often acquire the patents sold by other producing companies. A recent patent transaction that attracted media attention was the bankruptcy auction of the *Nortel Networks Corporation's* patent portfolio in June 2011. *Nortel*, a Canadian multinational telecommunications equipment manufacturer, filed for bankruptcy in January 2009. The patent portfolio, the most valuable asset of *Nortel*, went into auction. The patent portfolio consisted of ca. 6,000 patents and patent applications covering a wide range of technologies, including wireless, data networking, semiconductors, and Smartphone technologies. The patent portfolio was bought by *Rockstar Bidco*, a consortium of *Apple, Microsoft, EMC, Ericsson, Sony*, and *Research In Motion*, for USD 4.5 billion. The motives to buy these patents varied from acquiring essential patents for a certain standard to increasing the size of the patent portfolio for licensing negotiations, and to block competitors' access to the Smartphone related patents (Watson, 2010). Another example is the patent acquisition of *VisEn Medical*, a US-based producer of fluorescence in vivo imaging agents. In January 2010, *VisEn* acquired the fluorescence imaging agent IP portfolio and related technology platforms from *Bayer Schering Pharma*, a German pharmaceutical firm for an undisclosed amount. The acquired patent portfolio includes over 45 issued patents worldwide covering a wide range of fluorescence agent constructs and imaging methods. The main objectives of *VisEn* in acquiring this portfolio were to strengthen the patent position in in vivo fluorescent imaging agents, and to fill the pipeline of preclinical agent products and clinical imaging agents (Intellectual Property Today, 2010).

Besides producing companies that acquire patents based on defensive, financial, or strategic, and mainly intuitive and reasonable objectives, companies that do not produce goods have emerged as transaction partners in recent years. Even though they have none of the traditional acquisition motives, they are now significant players in the market for patents and technologies. For instance, in 2000 the *Golden Rice product development partnership* aggregated 11 patents from the agricultural companies *Syngenta, Bayer AG, Monsanto, Novartis, Orynova,* and *Zeneca Mogen* (Krattiger & Potrykus, 2007). The *Golden Rice product partnership* was founded to aggregate the patents and does not have any production or R&D. Another example that does not produce is *Intellectual Ventures*. In January 2009, *Intellectual Ventures* acquired the patent portfolio formerly developed and owned by *Transmeta Corporation* from *Novafora, Inc.,* a producer of digital video processors, for an undisclosed amount. The acquired patent portfolio contained more than 140 US patents and a substantial number of pending patent applications and some international patents and patent applications (Intellectual Ventures, 2009). Also in 2009, *Allied Security Trust*, a private company without production, bought 286 patents from the Japanese IT company *NEC*. The patents cover computer, graphics, microprocessor, and display technologies (Allied Security Trust, 2010). *Acacia Research* is another active patent buyer without production. For instance, in November 2011, *Acacia* acquired 65 US and foreign patents from the semiconductor manufacturer *Renesas* for an undisclosed amount (Wild, 2010b). These four examples show that even though these companies do not produce goods and therefore do not need patents in their historical meaning, they acquire patents and aggregate large patent portfolios. In the following sections, these companies are indicated as patent aggregating companies.

1.1.2 Practical challenges in leveraging corporate patent portfolios

Corporate leaders have recognized patents as a powerful instrument of corporate strategy (Cohen, Nelson, & Walsh, 2000; Davis, 2004; Grindley & Teece, 1997; Kash & Kingston, 2001; Rivette & Kline, 2000), and patents are no longer a legal matter but a strategic issue (Smith & Hansen, 2002). Companies have extended their patent departments to patent management divisions (Carlsson, Dumitriu, Glass, Nard, & Barrett, 2008), aligned patent strategy with corporate strategy (Cohen et al., 2000; Hall & Ziedonis, 2001), and now focus on leveraging their patent portfolios optimally (Davis & Harrison, 2001; Lichtenthaler, 2008b).

As patent management has received growing attention, the use of patents has developed from a primarily defensive and internal application (e.g., securing market shares by preventing competitors from entering the market, enforcing patents against infringers) to an active part of the company's strategy (e.g., licensing, sales of patents, external source of finance). From a patent management perspective, opening up the innovation process requires a shift from the traditional patent protection approach to a patent leverage approach using patents as means to exchange knowledge through selling and licensing. Therefore, literature distinguishes between internal patent exploitation and external patent exploitation (Kamiyama et al., 2006; Lichtenthaler, 2007c, Lichtenthaler, 2008a; OECD Publishing, BMWi, & EPO, 2005; de Rassenfosse, in press; Tietze, 2011). The internal exploitation of patents includes the protection of own products from copying or securing freedom to operate (Granstrand,

2000). Most companies have gained experience and able successfully to conduct the tasks of internal exploitation (Carlsson et al., 2008).

Firms increasingly exploit their patents externally. External patent exploitation occurs as licensing agreements, technology, or patent sales, or as a basis for collaborations with other companies (Birkenmeier, 2003; Ford, 1985; Shrestha, 2010; Vickery, 1988). Depending on the motives of the partners and on the characteristics of the exploited patents, the extent of external patent exploitation can vary. A patent can be licensed or sold in combination with (Anderson, 1979; Ford & Ryan, 1977; Marcy, 1979) or without (Lichtenthaler, 2007c; McDonough III, 2006; Shrestha, 2010) other technology knowledge of the firm.

Even though companies have realized that external patent exploitation is an integral part of leveraging patent portfolios, most companies still have major difficulties in conducting external patent exploitation projects successfully (Arora et al., 2001a). A recent survey proved that companies are willing to exploit 40% of their patent portfolio (on average) externally. However, until now, most of them have not been able to do so because transaction partners cannot be identified or transaction prices determined (Berneman, Cockburn, Agrawal, & Iyer, 2009). In contrast to product markets, the market for patents and technologies remains far from functioning well (Arora, Fosfuri, & Gambardella, 2001b; Caves, Crookell, & Killing, 1983; Cesaroni, 2004; Cesaroni & Mariani, 2001; Teece, 1981) because it lacks transparency regarding essential market information. Companies willing to trade are not able to gather information about buyers, suppliers, and technologies and patents offered. This lack of transparency in essential market information leads to high transaction costs (Arora et al., 2001a; Arora & Gambardella, 2010a; Caves et al., 1983; Ford & Ryan, 1981; Gambardella, 2002; Lichtenthaler & Ernst, 2007; Monk, 2009). Also, uncertainty regarding the quality of the patents (Gans, Hsu, & Stern, 2008; Troy & Werle, 2008), the value of the patents and the technology (Gambardella, Harhoff, & Verspagen, 2008; Scherer & Harhoff, 2000), and the transaction process (Lichtenthaler, 2004; Lichtenthaler, 2007a) prevent successful external exploitation of patents. Several empirical studies found that firms often still under exploit their patent portfolio (Elton, Shah, & Voyzey, 2002; Giuri et al., 2007; Rivette & Kline, 2000), implying that patents hold unused commercial potential.

As many firms are not able to overcome these market imperfections on their own, to find other corporate transaction partners and to exploit patents successfully, they seek help from service providers that are able to support or fulfill certain tasks. Therefore, a new business model has emerged: technology market intermediaries (Howells, 2006; Nambisan & Sawhney, 2007; Sapsed, Grantham, & DeFillippi, 2007). Technology market intermediaries may contribute to reduce market inefficiencies (Morgan & Crawford, 1996) through additional market knowledge in respect of technologies (Spulber, 1999), networks of potential transaction partners (Bryant & Reenstra-Bryant, 1998), and valuation experiences (Howells, 2006). Based on these competencies, technology market intermediaries could increase the number and the performance of external patent exploitation projects.

The number of technology market intermediaries is steadily growing. According to OECD Publishing et al. (2005), "market intermediaries have become more numerous and diverse as demand for technology transfer and patent valuation have grown" (p. 10). However, not all scholars advocate the emergence of technology market intermediaries and their benefits for producing companies. For instance, Lichtenthaler and Ernst (2008b) stated that, "the general facilitating role of intermediaries in technology transactions has to be questioned. Intermediary services have a positive effect on licensing revenues, but they do not significantly affect a firm's licensing performance relative to competitors" (p. 1025). Lichtenthaler and Ernst (2008b) recommended: "[...] firms need to develop internal resources for externally leveraging technology. Technology intermediaries should be regarded as a complement to internal activities, and they do not represent a substitute for internal resources" (p. 1027).

In summary, producing companies have recognized the shift from patents being only legal matters to serving as a strategic tool (e.g., Cohen et al., 2000; Davis, 2004; Grindley & Teece, 1997; Hall, 1992; Kash & Kingston, 2001; Rivette & Kline, 2000; Smith & Hansen, 2002). Therefore, most companies have established patent management divisions that fulfill the tasks of leveraging patent portfolios (e.g., Davis & Harrison, 2001; Lichtenthaler, 2008b). To leverage portfolios optimally, patents are internally (e.g., Arundel & Patel, 2003; Bader, 2006; Blind, Edler, Frietsch, & Schmoch, 2006a; Granstrand, 2000; Thumm, 2004) and externally (e.g., Cohen et al., 2000; Lichtenthaler, 2007b; Pitkethly, 2001; de Rassenfosse, in press; Rivette & Kline, 2000) exploited. In external patent exploitation transactions, the partners are other producing companies or patent aggregating companies. As producing companies still have difficulties in exploiting patents externally (e.g., Arora et al., 2001a;

Lichtenthaler & Ernst, 2009; Monk, 2009), and technology market intermediaries are limited beneficially (Lichtenthaler & Ernst, 2008a, Lichtenthaler & Ernst, 2008b), the question arises: are patent aggregating companies, as experienced buyers in the market for patents and technologies, an alternative approach for producing companies to leverage their patent portfolios optimally.

1.1.3 Deficits in current research

Patent aggregating companies reside within the context of patent management, market for patents and technologies, and intermediary literature. Literature on patent management refers to various fields of protecting innovations and inventions, for example, on patent strategy and on why firms acquire patents, or leveraging patent portfolios and external patent exploitation. Literature on the market for patents and technologies investigates market structures, market efficiencies, and players in the market. Literature on intermediaries refers to bridging companies that bring together supply and demand. Literature on the intersection of literature on intermediaries with literature on the market for patents and technologies deals with technology market intermediaries. These companies match supply and demand of patents, technologies, and innovation. Figure 1 illustrates the relevant literature streams and their connections. In the intersection of the three literature streams, patent aggregating companies and their utilization are located. Therefore, identifying deficits in current research publications from the three literature streams requires consideration.

Motivation 9

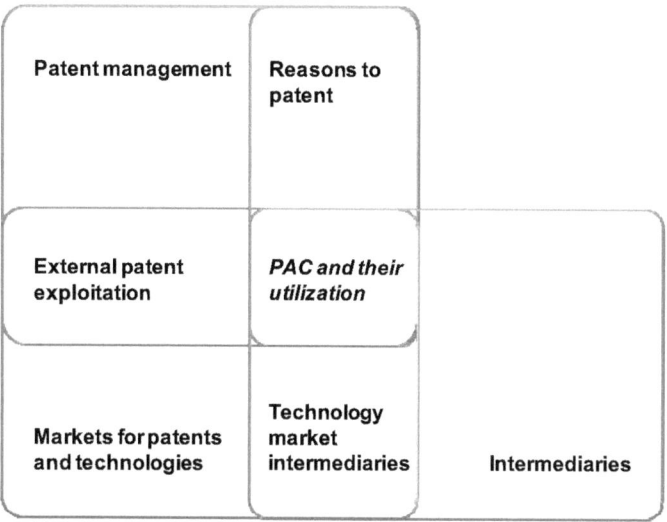

Figure 1: Relevant literature streams and research gap

Publications from the literature streams, patent management, market for patents and technologies, and technology market intermediaries, show several deficits regarding patent aggregating companies. These deficits are described in the following paragraphs.

During the last decade, the academic interest in external patent exploitation and the companies involved has increased, and the literature stream on the market for patents and technologies has developed. However, existing research is limited to the structure of the market, its inefficiencies, and producing companies or technology market intermediaries as main players in the market. The few studies recognizing patent buyers without own products classify these companies either as technology market intermediaries (e.g., Benassi & Di Minin, 2009; Monk, 2009; Tietze & Herstatt, 2010) or as threat in the market for patents and technologies (e.g., Chien, 2009; Geradin, Layne-Farrer, & Padilla, 2011; Golden, 2007; Johnson, Leonard, Meyer, & Serwin, 2007). Literature on non-corporate patent buyers is restricted to anecdotic evidence (e.g., Hetzel, 2010; Holden, 2011; Lipfert & Ostler, 2008; Pluvinage, 2011). Studies that approach patent aggregating companies without products but are active players in the market for patents and technologies in a more general and comprehensive manner do not exist.

In the literature on patent management, an extensive body of publications exists on the reasons why companies patent and acquire patents (e.g., Blind et al., 2006a, Blind, Edler, Frietsch, & Schmoch, 2006b; Chesbrough, 2003b; Cohen et al., 2000; Duguet & Kabla, 1998; Giuri et al., 2007; Hall & Ziedonis, 2001; Pitkethly, 2001; Shapiro, 2001). However, most of these studies focus on the motives of producing companies. Even though literature recognizes that companies without products but acquire patents, exist, little research has been conducted on them. Most studies analyze the motives of so called non-practicing entities, companies that own patents but do not have physical products and therefore, are non-practicing (e.g., Chien, 2009; Henkel & Reitzig, 2007; Lemley, 2007; Merges, 2009; Reitzig, Henkel, & Heath, 2007; Rubin, 2007). Several authors find that these companies buy infringed patents to enforce them (e.g., Ball & Kesan, 2009; Golden, 2007; Gregory, 2007; Henkel & Reitzig, 2007; Reitzig et al., 2007). Analysis of the motives of other patent acquiring companies without products that do not focus on infringed patents is scarce. The only study that exists in this area focuses on one company that acquires embryonic technology (Gredel, Kramer, & Bend, in press). A systematic and comprehensive analysis of the patent aggregating companies' motives to amass patents is lacking.

Literature on technology market intermediaries is fragmented and exists in academic and, to a certain extent, exists in non-academic literature. Academic, as well as non-academic publications mainly describe patent aggregating companies as technology market intermediaries. From a practitioner point of view, Millien and Laurie (2008) provide a collection of various IP business models. Based on the companies' self-description and on personal experiences as patent managers and patent management consultants, Millien and Laurie classify emerging and established IP business models in 17 different types, among them four different types of patent aggregating companies. In the academic literature, Benassi and Di Minin (2009) analyze patent brokers and their activities. They develop a typology of patent brokers that includes the identification of two types of patent aggregating companies. Other recent studies are limited to the distinction of defensive and offensive patent aggregating companies (Kelley, 2011; Pluvinage, 2011; Wang, 2010).

During the last decade, publications on non-practicing entities have emerged as a sub stream in the literature on technology market intermediaries. Most publications on patent aggregating companies are located in this sub stream and treat patent aggregating companies as non-practicing entities, companies that buy infringed patents and enforce them against large electronic companies. Luman III and Dodson (2006) describe the emergence of these companies and of how they affect innovation, companies, and society negatively. Reitzig et al. (2007) discuss the profitability of these companies and of how producing companies can counteract. Studies also analyze the impact of these companies on innovation (Shrestha, 2010), the market for patents (McDonough III, 2006), and on the characteristics of patents bought by these companies (Fischer & Henkel, 2009).

Within the sub stream of literature on non-practicing entities, patent aggregating companies are often connoted as either positive (Rubin, 2007) or negative (e.g., Henkel & Reitzig, 2008). Often patent aggregating companies are lumped together undifferentiated. Either they are suspected of acquiring patents only as litigation opportunity (so called 'patent trolls', e.g., Chien, 2009); or they are appreciated as white knight in underdeveloped markets for technology (so called 'patent elves', e.g., Geradin et al., 2011). Non-academic literature describes how customers of patent aggregating companies benefit from them or the influence they have on the patent market and other companies (e.g., Hetzel, 2010; Holden, 2011; Lipfert & von Scheffer, 2006; McCurdy & Reohr, 2008; Millien & Laurie, 2008; Pluvinage, 2011). A profound analysis of the business models and buying motives of patent aggregating companies, as well their interaction with producing companies has a limited availability in the literature stream of technology market intermediaries. The few studies focus mainly on the business model of non-practicing entities.

In summary, analyzing studies that explain the reasons for patenting, the reasons why corporate buyers acquire patents, and which business models in the IP sector exist and what they do, shows that existing literature is not able to explain where the differences in the business models of patent aggregating companies are, why these companies aggregate large patent portfolios, and how they have developed over time. Academic research on their activities and on how producing companies can utilize patent aggregating companies to leverage their patent portfolios does not exist. A comprehensive description and analysis of patent aggregating companies is lacking.

1.2 Research objectives and questions

This research is inspired by the practical need for assisting producing companies to leverage their patent portfolios. In this respect, it has been argued that many companies have started to exploit patents externally but still are not able to overcome the problems in the market for patents and technologies. As patent aggregating companies have emerged as significant and experienced buyers in the market for patents and technologies, producing companies may use them to leverage their patent portfolios. Hence, little is known about patent aggregating companies, and they have not been addressed sufficiently in the existing literature to date. The main objectives of this study are to shed light on patent aggregating companies and their business models, and to develop a management framework that helps producing companies to leverage their patent portfolios optimally. Thus, this research aims at answering the following question (Figure 2):

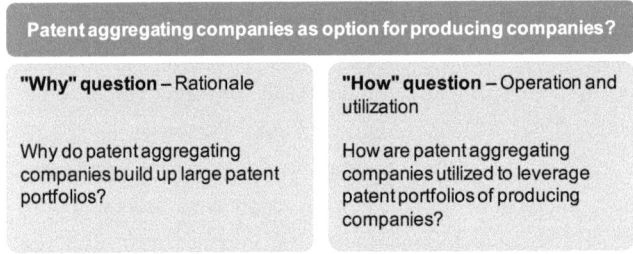

Figure 2: Research questions

To provide answers to these questions, this research focuses on the analysis of the patent aggregating company itself. Analyzing patent aggregating companies, their business models, strategies, and activities comprehensively, this study develops results on how patent aggregating companies can generate benefits for producing companies. Thus, this research contributes to existing theory and literature on the market for patents and technologies by shedding light on new and important players, the evolvement of these players, and by developing a typology that conceptualizes these players. Furthermore, this research aims at translating empirical and theoretical insights from patent aggregating companies into managerial relevant practice by providing a management framework for producing companies on how patent aggregating companies can be utilized for leveraging patent portfolios. Thus, the study moves beyond the existing literature by deriving deep insights and managerial recommendations in contrast of merely treating it as an empirical phenomenon.

1.3 Terms and definitions

The term patent aggregating companies and its concept is not a frequently cited term in the literature. To distinguish patent aggregating companies from other companies such as producing companies, technology market intermediaries, and sub-groups of patent aggregating companies the following definition describes patent aggregating companies in this research:

Patent aggregating companies are firms that focus on amassing patents, see R&D not as a core competency, and do not produce or manufacture own physical goods.

Aggregating patents is used as the general term for acquiring ownership or commercialization rights on a patent. As patents can be exploited, monetized, advanced, or defended without ownership rights, some patent aggregating companies conclude contracts with the original patent owner to exploit/commercialize/monetize the patents exclusively, and ownership of the patent does not change to the patent aggregating company. Acquiring a patent and transferring the ownership to the patent aggregating company is a permanent and irrevocable action and cannot be limited. In the following research, the term *amassing and aggregating* is used as a synonym. In cases where the difference between acquiring ownership rights and acquiring commercialization rights is important for the analysis of patent aggregating companies and their strategies and business models, it is mentioned explicitly.

The above-stated definition does not quantify the number of patents a patent aggregating company amasses. As patent aggregating companies are a young phenomenon and the business models are emerging and vanishing quickly, the number of patents they aggregate is heterogeneous. In this research, companies that have amassed 10 patents or more and have patent aggregation as their main business model are designated as patent aggregating companies.

In addition, companies that focus on R&D are excluded. Especially in the high-technology industry, several companies exist that account large R&D expenditures and at the same time acquire patents but do not have their own production. These companies create an ongoing stream of innovation and use patents to protect their innovation. The developed technology is not used to produce goods, but the patents are licensed to operating companies that manufacture products and employ the technology. Laurie (2006) defines these types of companies as IP factories.[1] IP factories are excluded from the term patent aggregating companies.

A patent aggregating company acquires patents (and ownership rights) or commercialization rights from the original patent owner (supply side of the patent aggregating company). The patent aggregating company may sell the patents to a new patent owner or out-license the patents to a licensee (demand side of the patent aggregating company). Figure 3 illustrates the relation between a patent aggregating company and its supply and demand side.

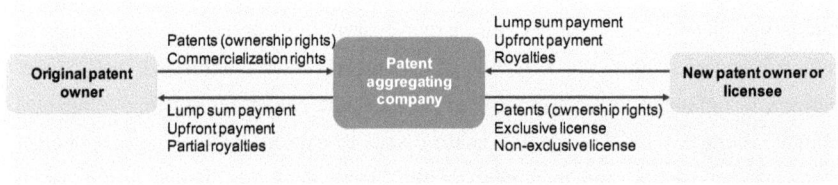

Figure 3: Players and relationships in the patent aggregating ecosystem

The original patent owner can be a natural person, such as a single inventor, or a research institution, a university, or a corporate entity, such as a small and medium enterprise (SME) or a multinational enterprise (MNE). Either the original patent owner can be the inventor or have acquired the ownership rights from third parties.

---

[1] One example for this type of companies is *Tessera Technologies*, an US based company that invests in, licenses, and delivers innovative miniaturization technologies for next-generation electronic devices. In the year ended on December 31, 2009 *Tessera* had R&D and other related costs of USD 65.9 million by total revenues of USD 299.4 million. Beside generating patents, *Tessera* also acquires patents. A recent example for *Tessera's* acquisition activities is the purchase of 64 patents and patent applications from *ALLVIA*, a US-based Through-Silicon Via development company (Tessera Technologies, 2011).

For transferring ownership rights or commercialization rights to the patent aggregating company, the original patent owner is remunerated with different compensation elements, such as lump sum payments, upfront payments, or royalties in a profit-sharing model. In general, royalties are paid periodically and are often tied to performance targets, such as annual revenues. The final compensation elements depend on the contractual agreement. In certain cases, additional compensation elements, such as payments for R&D and service contracts, or back-licenses to the original patent owner for specific fields of use are possible.

The new patent owner or the licensee can be a natural person, such as a single inventor, a research institution, or university, or a corporate entity, such as a SME or MNE. The new patent owner buys patents and with it, ownership rights from the patent aggregating company and pays, depending on the contractual agreement, a lump sum payment or upfront payment and royalties. A licensee takes an exclusive or a non-exclusive license from the patent aggregating company and pays, depending on the contractual agreement, royalties or an upfront payment and royalties. A license is a legal contract where the patent aggregating company grants exploitation rights over a patent to a licensee. With an exclusive license, the licensee has the sole exploitation right. Typically, a license is granted for a limited period of time and to specific industrial or geographical markets.

1.4 Research concept and methodology

The awareness regarding the exploitation of patents, as well as technology market intermediaries has tremendously increased during the last decades. Patent aggregating companies have emerged as new empirical phenomenon and empirical insights in this type of companies are scarce. This research aims to contribute to existing literature on the market for patents and technologies, technology market intermediaries, and external patent exploitation by building representation of observable elements, generating questions, and presenting propositions relevant to explaining phenomena (Eisenhardt, 1989a; Kromrey, 1998). In inductive research, new findings are derived from existing literature, as well as the insights that originate directly from the data analysis. The data are generated in field research. Theory is built through connecting and disconnecting data and existing literature throughout the entire research process

(Mintzberg, 2007), so that sufficient depth is provided to achieve an understanding of these interrelations and dynamics (D'Iribarne, 1996).

Research methodology

Patent aggregating companies have emerged as a very young empirical phenomenon. Due to the novelty and complexity of the topic, this study applies an exploratory, qualitative research design. Therefore, case study research according to Eisenhardt (1989a) and Yin (2009) is employed. Case study research allows an understanding the phenomena under investigation while addressing detailed questions to gain deeper insights into patent aggregating companies' strategies, motives, and activities (Yin, 2009).

More precisely, the research follows a multiple-case approach with the patent aggregating company as the unit of analysis. A multiple-case approach allows for cross comparison. These cross comparisons capture the specific aspects of all sites and set the aspects in their natural perspective.

In qualitative research, the analytical process is an iterative one. Constant alternations between data collection and data analysis are accomplished. The data analysis of this research follows Eisenhardt's (1989) approach to building theory from case study research. A reference framework is built from existing literature and theoretical insights relevant to explaining the phenomenon are the basis of the phenomenon's exploration (Eisenhardt, 1989a; Miles, 1979). The reference framework selects and explains the main aspects to be studied within the case studies (Voss, Tsikriktsis, & Frohlich, 2002). The derivation of the framework from literature ensures that the subsequent collection of qualitative data is based on a sound theoretical approach. In addition, it guides the data collection. Throughout the study, mini-cases and narratives are used to illustrate theoretical concepts for the potentials of patent aggregating companies and of how to realize them.

The main criteria in qualitative empirical research are the reliability and validity of results. Usually, three types of validity can be differentiated: construct validity, internal validity, and external validity. According to Yin (2009), in this research, construct validity is ensured by using multiple sources of evidence and establishing a chain of evidence between the questions asked, data collected, and conclusion drawn. To ensure internal validity in this research, three strategies are employed. The collected data comes from multiple sources (Lamnek, 1995; Yin, 2009). Semi-structured interview data are combined with the results of thoroughly conducted desk research, internal documentation, and presentations by experts and management. In addition, member checking is conducted. To determine the accuracy of the qualitative findings, the informants serve as a check throughout the analysis process in an ongoing dialog. Peer examination was conducted with the peer examiners Professor Gassmann (ITEM-HSG, University of St. Gallen), Professor Webster (Melbourne Institute of Applied Research and Social Science, University of Melbourne), and the faculty of the research group 'Industrial Organization' of the Melbourne Institute of Applied Research and Social Science. To ensure external validity, the primary strategy is to provide a detailed description of the research and to set up a detailed case study protocol and database (Yin, 2009). This strategy allows anyone interested in testing transferability to compare results (Merriam, 1998). Reliability is to ensure that another researcher could repeat the research with the same procedure (Eisenhardt, 1989a). Therefore, in this study, data collection and analysis are described in detail to increase transparency (e.g., the researcher's role, the informant's position, case selection criteria, context of data collection). Furthermore, the triangulation of data strengthens not only the internal validity, but also supports reliability.

Research sample

Current research offers little information regarding motives, strategies, and activities of patent aggregating companies. Therefore, several case studies have been selected and studied in detail to gain an in-depth understanding of their natural setting, complexity, and context (Punch, 2005). The research was carried out in two phases during 2009 and 2011.

The first research phase included interviews with corporate patent experts and non-corporate patent experts to explore technology market intermediaries and patent service provider in general. It is based upon 93 interviews with 68 patent service providers and technology market intermediaries predominantly based in Europe and the US. The interviews stems from seminar works, scientific industry studies, and participated contracted research projects at the Institute of Technology Management at the University of St.Gallen under the supervision of the author. All companies are engaged with patent transactions. The various company contacts represent a wide range of industries, technology categories, and business models. The inter-industry scope of companies represents the heterogeneity of the explorative phase.

In the second research phase, an in-depth analysis of companies with diverse but distinctive business models was conducted. Based on the case firms investigated in phase one, the selected firms have the highest potential for learning new insights with respect to their business models, strategies, and activities, as the firms have been proven to be successful in the intransparent market for patents. Thus, sampling has been conducted according to theoretical rather than random sampling (Eisenhardt, 1989a). While random sampling is typically found in theory testing on a broad scale, theoretical sampling is the preferred sampling strategy when new or existing theory is developed or advanced (Eisenhardt, 1989a; Yin, 2009). In qualitative research, there is no ideal number of case studies. Due to the broad spectrum of business models, this research presents in-depth case studies of 27 firms that show very distinct strategies and high activity in the market for patents.

Table 1 shows the analyzed companies and therefore, the empirical base of this research.

Research Phase	Number of Interviews	Companies
Phase I Literature review and explorative interviews on external patent exploitation and technology market intermediaries	93	1790 Capital Management, 5i Principles Group, 5iTech, Acacia Research, Alliacense, Allied Security Trust, Alpha Gasser Patentverwertungs AG, Alpha Patentfonds Management GmbH, Altitude Capital, Anadeus Ltd, Blueprint Ventures, Burford Group Limited, Capital Royalty L.P, Caisse des Depots, Chipworks, Coller Capital, CONSOR, CreativE[1], Credite Suisse, Deutsche Bank, Drakes Bay Company, European Investment Funds, Fergason Licensing, Finance Systems, Gathering 2.0, General Patent Corp, ICAP Ocean Tomo, IgniteIP, Inflexion Point, Innovaro, Intellectual Ventures, IP Auctions GmbH, IP Bewertungs AG, IP Exchange Chicago, IP Navigation Group, IP Trade, IPEG Consultancy BV, Juris Capital, Kratos Ventures, Marqera, MPEG LA, New Venture Partners, NineSigma, NW patent funding, Ocean Tomo, Ocean Tomo Indexes, Open Invention Network, Papst Licensing, Paradox Capital, Patent board, PATEV, PCT Capital, Pete Invest MedTech[2], Plutus IP, Rambus, Rembrandt IP, Royalty Pharma, RPX, Sisvel, Steinbeis TIB, TPL, Techquity, Thinkfire, Tynax, UBM TechInsights, Via Licensing, Wi-LAN, Yet2.com
Phase II In-depth case studies of patent aggregating companies	44	Acacia Research, Allied Security Trust, Alpha Gasser Patentverwertungs AG, Alpha Patentfonds Management, Capital Royalty, Coller Capital, CreativE[1], Deutsche Bank, Fergason Licensing, Finance System, General Patent Corporation, IgniteIP, Intellectual Ventures, IP Bewertungs AG, IP Navigation Group, MPEG LA, Open Invention Network, Papst Licensing, Patent Freedom, PATEV Associates, Pete Invest MedTech[2], RPX, Steinbeis TIB, Techquity Capital
Total	137	68

[1] The name of the company has been disguised for confidentiality reasons. In this research, the company is referred to under a fictitious name. The name *CreativE* replaces the firm's actual name.

[2] The name of the company has been disguised for confidentiality reasons. In this research, the company is referred to under a fictitious name. The name *Pete Invest MedTech* replaces the firm's actual name.

Table 1: Empirical base of research

The selection procedure was part of the iterative interview process within the first phase. The selection criteria for the case studies included a focus on pioneering activities in the market for technologies and patents. In addition, the selected companies had to amass patents rather than only transfer them or consult patent sellers. In addition, the selected case study firms represent the two main regions where patents and technologies are traded: Europe and the US. The case study selection focused on firms that aggregate patents from industries with a high propensity to patent and a high relevance of patents as a high-technology industry or a life science industry. As patent aggregating companies are very heterogeneous, companies of all sizes were selected.

Location	Focus on acquisition of patents from industry:		
	Electrical engineering	*Life science*	*No industry focus*
Headquarters USA	Acacia Alliacense Allied Security Trust Fergason Patent *IP Holdings* *MPEG LA* RPX Sipro Lab Open Inv. Network Via Licensing	Capital Royalty Pete Invest MedT[1] Royalty Pharma *IP Holdings* *MPEG LA*	AlseT Coller Capital IgniteIP Intellectual Ventures IP Navigation Rembrandt Techquity
Headquarters Europe	CreativeE[2]	Golden Rice PDP	Alpha Patentfonds Eco-Patent Commons Papst Licensing Patent Invest Fond Patent Select

[1] The name of the company has been disguised for confidentiality reasons. In this research, the company is referred to under a fictitious name. The name *Pete Invest MedTech* replaces the firm's actual name.
[2] The name of the company has been disguised for confidentiality reasons. In this research, the company is referred to under a fictitious name. The name *CreativE* replaces the firm's actual name.

Table 2: Research sample of patent aggregating companies

About more than two-thirds of the in-depth case studies are from US-based patent aggregating companies (see company list Table 2). Detailed case studies are described for *Alpha Patentfonds, Intellectual Ventures, Pete Invest MedTech* [2], *Patent Select, Acacia Research Corporation, Allied Security Trust, MPEG LA, and Golden Rice product development partnership* while mini cases provide examples throughout the text. These emphasize differences in strategies and motives, as well as in the activities of patent aggregating companies.

Data Collection

In all phases, data has been collected through personal face-to-face or telephone semi-structured interviews of 45–90 minutes in length. The main interview partners were typically founders, CEO, CPO, CFO, partners, managing directors, or vice presidents of patent aggregating companies. Informants in top management positions were interviewed mainly to secure the information on the strategic direction and the activities of the patent aggregating companies. Some of the respondents have been interviewed more than once for follow up questions and approval of earlier data. To ensure consistency, the same semi-structured interview guide has been used throughout all interviews. Whenever possible, this interview guide has been sent to the interviewee in advance. All interview data is complemented by written company information, such as internal memorandums, presentations, and publicly available information to increase validity. In addition, follow-up interviews have been conducted to confirm the case study interpretations from the interview data. This triangulation through combining multiple sources of evidence contributes to confirming the validity and reliability of the research data (Voss et al., 2002; Yin, 2009).

1.5 Thesis structure

This thesis is structured as follows (also see Figure 4):

Chapter 2 reviews the relevant literature regarding patent management, internal and external patent exploitation, and patent intermediaries. It provides the basic understanding of patent management and of how producing companies leverage their

---

[2] The name of the company has been disguised for confidentiality reasons. In this research, the company is referred to under a fictitious name. The name *Pete Invest MedTech* replaces the firm's actual name.

patent portfolios. Drawing on the reviewed literature, a reference framework for the analysis of patent aggregating companies is developed. To gain deeper insight into patent aggregating companies, the reference framework uses four defining characteristics: setting, strategy, organization, and process.

As patent aggregating companies are still a black box, *Chapter 3* explores the empirical phenomenon and provides insight into the general setting of patent aggregating companies, their history, and the venture's funding. The process of patent aggregation is analyzed and decomposed into four phases, and the strategies of patent aggregating companies identified. Based on the empirical data, eight business models of patent aggregating companies are distinguished.

Chapter 4 analyzes the potentials patent aggregating companies offer producing companies. Patent aggregating companies can create opportunities in the environment of producing companies, as well as within the company. Therefore, external and internal potentials are distinguished.

Chapter 5 identifies four different archetypes of patent aggregating companies: (1) the merchant, (2) the gardener, (3) the collector, and (4) the patron. The archetypes differ regarding their competencies and the rewards they provide to the original patent owners. Two different business models represent each archetype, and their characteristics are illustrated with eight cases studies.

Chapter 6 provides recommendations how producing companies can utilize patent aggregating companies to leverage their patent portfolios. Based on value creating options and on constraints in the utilization of patent aggregating companies, a management framework is developed. The patent aggregating business, as well as the market for patents and technologies, are highly dynamic conditions and change fast. As the management framework reflects the current situation of patent aggregating companies, the development of patent aggregating companies and the main driver of the development are analyzed.

Chapter 7 concludes this thesis and presents implications of this study from both a theoretical and an empirical point of view. The outlook describes further research and trends.

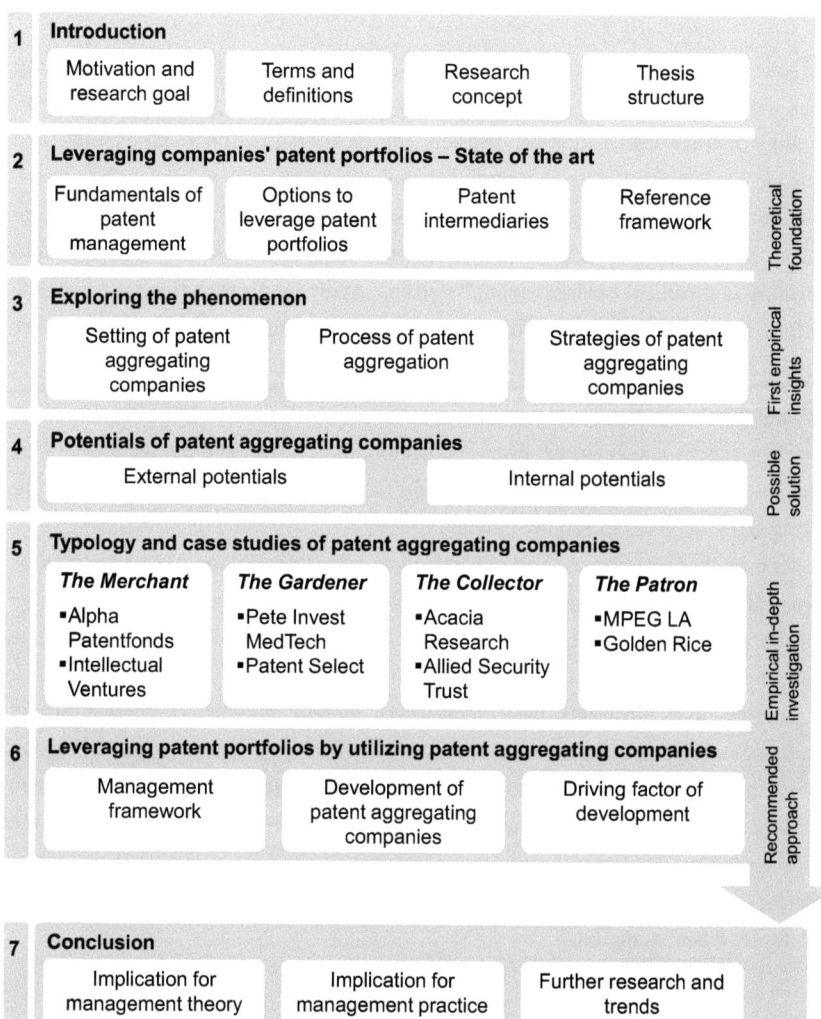

Figure 4: Structure of the thesis

2 Leveraging companies' patent portfolios – State of the art

Based on the derived research objectives (section 1.-1313864729), the following chapter provides a review from the literature streams of patent management, technology market intermediaries, markets for technologies, open innovation, and knowledge management. The last part of this chapter derives a reference framework from literature. This reference framework serves as a guideline for the data collection and data analysis and it is defined through the four key characteristics of patent aggregating companies' business models: setting, strategy, organization, and process. Therefore, the literature is reviewed and presented according to its relevance to answer the research questions and to build the reference framework:

(1) Literature that analyzes *why firms patent or acquire patents* from the literature streams patent management, open innovation, and knowledge management. This analysis serves as theoretical basis for the analysis of patent aggregating companies' motives to acquire patents.

(2) Literature that describes *how producing companies leverage their patent portfolios* from the literature streams of patent management and open innovation. It serves as the basis to understand for which opportunities producing companies could utilize patent aggregating companies.

(3) Literature that illustrates *what technology market intermediaries are, which activities they perform*, and *non-practicing entities* as a controversial sub-group from the literature streams of technology market intermediaries. It provides a basic understanding of the business models, the recent discussions, and the differentiation of patent aggregating companies. Further, it shows that besides patent aggregating companies' strategies, operational factors distinguish the different business models.

2.1 Fundamentals of patent management

The growing number of patent applications over the last century underlines the trend that patents have become an important corporate asset and a critical driver for business profitability in the knowledge economy (Teece, 2000) is underlined by the growing number of patent applications over the last century. Since 1985, the worldwide yearly patent filings have more than doubled. The World Intellectual Property Organization

(WIPO, 2010) reports that 1.9 million patent applications were filed worldwide in 2008. The major patent application countries are the US with 456,321 patent applications, followed by Japan with 391,002 and China with 289,838 patent applications. At the European Patent Office (EPO), 146,150 patents were filed in 2008. Thus, more and more firms have started to accumulate patent rights. The following part presents the reasons why firms patent and acquire patents.

2.1.1 Reasons why firms patent

The intention of the patent system is that firms can exclude third parties from using their invention and therefore, are able to appropriate returns from innovation (Levin, Klevorick, Nelson, & Winter, 1987; Teece, 1986). Therefore, the traditional motive for patenting is to use patents to protect innovations from imitation and to secure freedom to operate (FTO). Firms apply for patent protection to improve their competitive position (Hanel, 2006). However, the reasons to patent have changed and have become more complex and comprehensive (Blind et al., 2006b). Concomitant, the relative relevance of the traditional protection motive has been reduced (Blind et al., 2006a).

An increasingly important motive to patent is to block competitors (e.g., Blind, Cremers, & Mueller, 2009; Thumm, 2004). Duguet and Kabla (1998) observe that protection of own innovations from imitation and blocking competitors are the two central motives why firms patents. Blind et al. (2006a) differentiate between offensive blockade, a firm patents to prevent other firms from using their own inventions in the application filed of the patenting company, and defensive blockade, a firm patents to prevent being restricted in its own technological field. Empirical studies identify two more defensive patenting motives. Firstly, companies patent to prevent competitors from developing around their technology (Arundel & Patel, 2003). Secondly, companies patent to build up large portfolios to prevent patent infringement lawsuits or to improve their own negotiation position in cases of litigation. In this case, patents are generated as bargaining chips in negotiations (Cohen et al., 2000; Granstrand, 2000; Hall & Ziedonis, 2001; Noel & Schankerman, 2011).

In addition to the defensive motive, companies patent nowadays for a variety of reasons. Increasingly, companies file patents to have a neutral R&D controlling instrument that serves as indicator to assess and reward R&D personnel (Blind & Thumm, 2004) and to internal evaluate the R&D productivity (Arundel, van de Paal, &

Soete, 1995). During the last decades, patents are filed to pursue market strategies, and companies patent to access international markets (Duguet & Kabla, 1998) or to create a tool for reputation management in external evaluations with strategic or financial partners (Cohen, Goto, Nagata, Nelson, & Walsh, 2002). Companies are increasingly monetary motivated and file patents to generate licensing revenues (Pitkethly, 2001) or additional cash flows from patent sales (Rivette & Kline, 2000) and to access other forms of company funding (de Rassenfosse, in press).

The motives to patent differ between industries. Cohen et al. (2000) argue that the number of patentable elements in a commercializable new product is important in affecting the reasons why companies use patents and hence, why they patent in the first place. They distinguish between 'complex' versus 'discrete' product industries. Industries with products that comprise many separately patentable elements are 'complex product industries'. Examples for complex product industries are high-technology industries, such as the medical devices industry, semiconductor industry, or telecommunication industry. Industries with products that comprise only one or very few separately patentable elements are 'discrete product industries'. Examples are the chemical or pharmaceutical industry.

Companies operating in complex product industries rarely have full control over all essential patents they need to produce their products. Several competitors have patents that all players in this technological field need. Therefore, all players are depending on each other and use patents as exchange potential or negotiation material. Various studies confirm that the share of cross-licensing in high-technology industries is above average (Anand & Khanna, 2000; Giuri et al., 2007; Grindley & Teece, 1997; Hall & Ziedonis, 2001). As patents have to be shared and transferred, they become a currency, and patent thickets[3] arise in complex product industries as an exchange forum for complementary technologies (Reitzig, 2004b). Companies operating in discrete product industries are able to make products with full control over all essential patents. Therefore, they use patents to secure market access, to block competitors, and

---

[3] Defined by Shapiro (2001), patent thickets are "a dense web of overlapping intellectual property rights that a company must hack its way through in order to actually commercialize new technology" (p. 119). Patent thickets arise mainly in complex industries, such as the medical devices industry, semiconductor industry, or telecommunication industry. Literature argues that patent thickets may raise transactions costs that arise from circumvention and long negotiations and contracting between players that increases costs over the positive impact on R&D incentives (Cockburn, MacGarvie, and Müller, 2010).

to secure freedom to operate (Kash & Kingston, 2001; Roberts, 1999). However, not all studies support this result. Blind et al. (2006a) did an analysis of 522 German companies and could not observe differences in the reasons to patent between complex and discrete product industries. Most literature generally agrees on the fact that patents in the chemical or pharmaceutical industry are an important and effective means to protect innovations (e.g., Cohen et al., 2000; Levin et al., 1987; Mansfield, 1986).

Besides the general motives to patent and the differences in the industries literature's discussion of patenting behavior and of why firms patent focuses on geographical regions, especially the US, Europe, and Japan (e.g., Carlsson et al., 2008; Cohen et al., 2002; Ernst, 1995; Granstrand, 2000, Granstrand, 2004; Pitkethly, 2001). So far, all publications focus on the reasons why companies that produce goods patent.

2.1.2 Reasons why companies buy patents

Apart from their own patenting activities, companies can also build up patent portfolios by acquiring patents. Compared with the amount and intensity of studies on the reasons why firms patent, the reasons why firms buy patents is analyzed to a much smaller extent and often only as a by-result. Literature streams that observe the motives for acquiring patents are mainly studies on open innovation and knowledge acquisition. Since Chesbrough's seminal work (2003a), literature on open innovation is burgeoning, and patents play a crucial role because they are the legal basis of the in and out flows of knowledge (e.g., Arora, 1995; Chesbrough, 2003b; Chesbrough, 2006; Lichtenthaler, 2007b). The open innovation process encompasses inbound, outbound, and coupled activities (Gassmann, 2006). Acquiring knowledge and patents from external sources is part of the inbound activities (Chesbrough, 2003a). The reasons why firms acquire patents in the open innovation process are mainly technological and help companies benefit from external innovation and R&D activities (e.g., Cohen & Levinthal, 1990; Grindley & Teece, 1997; von Hippel, 1988; Pisano, 2006; van de Vrande, Vanhaverbeke, & Gassmann, 2010; Westney, 1993). The technological motives to acquire patents include the objective to fill the development pipeline and acquire new ideas (Nambisan & Sawhney, 2007); to allow for more varied product development (Atuahene-Gima, 1992; Cesaroni, 2004); to enter new markets (Contractor, 1980); to establish new products or to go into a new line of business (Cesaroni, 2004; Jones, Lanctot, & Teegen, 2001); to reduce R&D costs

(Hoffmann & Schlosser, 2001; Mohr & Spekman, 1994); and to reduce the risks of R&D failures (Reepmeyer et al., 2011).

In addition to technological reasons, defensive reasons are increasingly important for companies. To guard against the risk of patent litigation, companies acquire patents so they can strike back against or neutralize threats of enforcement lawsuits brought by their competitors (e.g., Chien, 2009; Ziedonis, 2004). Therefore, the defensive reasons to acquire patents are quite similar to several reasons of why firms patent. Also the motive to build up large portfolios that are used as bargaining chips against other patentees, competitors, or suppliers, and therefore, can prevent enforcement lawsuits is important (e.g., Hall & Ziedonis, 2001; Williamson, 1983). Buying patents and taking the patents from the open market eliminates the chance of litigations (Ziedonis, 2004). Thus, it reduces not only the threat of litigation, but for the acquiring company, it also generates the freedom to operate (Shapiro, 2001). So far, studies focus on the reasons why companies that produce goods acquire patents.

2.2 Options to leverage patent portfolios

As the number of patent applications grows and with them the size of companies' patent portfolios (Blind et al., 2009; Hall, 2005; Hall & Ziedonis, 2001), firms have established dedicated resources to manage patent portfolios (Bianchi, Chiaroni, Chiesa, & Frattini, 2011b; Grindley & Teece, 1997). Most companies now leverage patents as an integral part of business strategy (e.g., Arora et al., 2001b; Cohen et al., 2000; Davis, 2004; Rivette & Kline, 2000). Patents can be leveraged in two different ways, either through internal or external exploitation (Benassi, Corsaro, & Geenen, 2010). For the many companies leveraging patents externally, it is still a difficult and often unsuccessful task because they face several impediments to external patent exploitation (e.g., Arora et al., 2001a; Caves et al., 1983; Gans & Stern, 2010; Guilhon, 2004). The following part describes the different options a company has to leverage its patent portfolio. Secondly, the impediments to external patent exploitation and technology transfer are illustrated.

2.2.1 Internal and external exploitation of patents

According to Kelley (2011), patents can generate value from three different perspectives: financial, strategic, and defensive. Patents create financial value through monetary benefits and cash flows that directly result from transactions of the patents and include, for instance, received royalties or sales proceeds. The strategic value of patents is represented by the impact a patent has in the product market. In addition, patents create a defensive value by avoiding litigation and the costs of patent litigation, for instance, searching for circumvention in the case of injunction (Kelley, 2011).

To extract the three forms of value, the literature distinguishes two ways in which a company can leverage the patent portfolio: either through internal or external patent exploitation (Benassi et al., 2010; Lichtenthaler, 2005). Whereas in the past, studies discussed that the two ways can only be conducted exclusively, studies increasingly discuss that external exploitation can, and in several cases should be conducted simultaneously to internal exploitation (e.g., Lichtenthaler & Lichtenthaler, 2009; Mathews, 2003). In some cases, the internal exploitation of patents is a prerequisite for a successful external exploitation (Arora et al., 2001b). In other cases, the external patent exploitation is precondition for a successful product business (Koruna, 2004; Lichtenthaler, 2006).

The traditional way of patent exploitation, resulting from the nature of the patent system, is internal exploitation. The economic rationale behind the creation of a patent system is that inventors are rewarded with a temporary monopoly from their invention for disclosing their invention to the public (Gassmann & Bader, 2011). The EPO (2009) defines a patent as "a legal title which protects a technical invention for a limited period. It gives the owner the right to prevent others from exploiting the invention in the countries for which it has been granted. All patents are published, so everyone can benefit from the information they contain" (p. 5). Patents are negative rights and do not allow the owner the exploitation of the patented invention, but exclude third parties to sell or use the invention. On the one hand, society benefits because the patent discloses an invention that promotes the state of the art. On the other hand, this exclusion of others helps the inventor to monopolize his/her rewards from R&D. Therefore, patents generate a strategic value by protecting innovation from imitation or copying.

Often it is not possible to sustain the competitive advantage with only one patent. Therefore, patents can create strategic value by preventing circumvention (e.g., Arundel & Patel, 2003; Ball & Kesan, 2009), by blocking the company's core technology and potential substitution technologies (e.g., Hanel, 2006; Thumm, 2004), and erecting entry barriers for market access (e.g., Caves, Whinston, & Hurwitz, 1991;

Gilbert & Newbery, 1982). In addition, to strategic value, internally exploited patents generate defensive value and ensure freedom to operate for the patent owner (Lichtenthaler, 2011). Freedom to operate ensures that R&D and the production of goods can be conducted without interfering with third parties' rights. Often, the exclusive internal exploitation of patents is mainly considered for protecting the firm's technological core competencies.

Patents are externally exploited in the secondary market for patents, also called market for patents and technologies (e.g., Arora et al., 2001a; Gambardella et al., 2007; Svensson, 2007). The market for patents and technologies is a broad term and denotes trade in technology disembodied from physical goods (Arora & Gambardella, 2010b). To exploit patents externally, the patent owning company has to decide between assigning or licensing patents to a third party (Benassi et al., 2010).

By assigning patents to the other transaction party, the original patent owner transfers the ownership rights to the new patent owner. By licensing patents to the other transaction party, the original patent owner authorizes the licensee to use the patent, patents, or patent portfolio. The patent owner can assign or license either only the sole legal right of exclusion (McDonough III, 2006) or the right or exclusion in combination with additional technology or knowledge (Marcy, 1979).

The classic option of assigning the ownership rights to another party is selling patents. By selling the patents, the patent owner can generate financial value and realize the remaining value from unused inventions and patents. These patents can result from terminated research projects, cover technology that lies outside the core competency of the company, be remnants from a shift in business strategy, acquired in corporate M&A transactions, and they are now unrelated to products or cover technology in a different stage of the actual product life cycle (Lichtenthaler, 2005). Patent sales also generate financial value through costs reduction. Instead of abandoning patents that are not practiced, the sale of patents can save renewal fees beside the onetime cash inflow through the selling price. One successful example for financial value generation from costs savings is *Dow Chemical*. In the 1990s, *Dow* changed its patent strategy. Instead of abandoning patents, *Dow* could sell its unused patents. Beside the acquisition price, this saved more than USD 50 million (Davis & Harrison, 2001).

Patents can also generate strategic value if they are transferred together with technology. In some cases, the patent owners develop an invention but face constraints that impede the commercialization of the innovative product (e.g., Bianchi, Chiaroni, Chiesa, & Frattini, 2011a; Zahra & Nielsen, 2002). To overcome these constraints, companies can sell patents, technologies, and knowledge to a third company (Gredel et al., in press). On the one hand, this transaction generates cash flows that can be reinvested in R&D or commercialization efforts of the sold or another technology (Lipfert & Ostler, 2008). On the other hand, through a grant back license and the additional resources of the transaction partner, the original patent owner and the new patent owner can commercialize the innovative technology jointly.

Another option to generate financial value from patents is to create a spin-off and transfer patents and technologies to this new company (Chesbrough, 2003c; Davenport, Carr, & Bibby, 2002). An example for a successful spin-off is *Actelion*. Roche restructured its cardiovascular therapy area and abandoned the *Bosentan* project in 1997. After that, four managers from this project founded the company *Actelion*, and *Roche* assigned the patents that cover *Bosentan* to *Actelion*. *Actelion* started the clinical development phase 3, changed the original indication of *Bosentan*, and brought the drug to the market. Therefore, *Roche* received cash flows from a project that the company originally stopped, and *Actelion* could commercialize a new drug without early development risks (Reepmeyer, 2006).

Another option to generate value from patents is to donate patents (Bader, Gassmann, Ziegler, & Ruether, in press). Donating patents to a non-profit organization generates financial value not through direct cash inflows but through tax savings (Carlsson et al., 2008). Donating patents to an open source community generates strategic and defensive value. The basis of open source communities is that participants give access to inventions free of charge so that all other participants can build on each other's innovations (Yanagisawa & Guellec, 2009). Open source is particularly relevant in software development (von Krogh & von Hippel, 2003) but increasingly, companies from other industries become aware of its potential (e.g., developing a cure for tuberculosis in developing countries (Dhanaraj, in press). By participating in open source projects and donating patents, companies can generate strategic value through developing business models that sell complementary services to open source products (e.g., Cusumano, 2004; Dahlander, 2005; Fitzgerald, 2006; Watson, Boudreau, York, Greiner, & Wynn, JR., 2008) or through developing products with the innovative power of outsiders, which is offered to the company free of charge (e.g., Andersen-

Gott, Ghinea, & Bygstad, in press; Dahlander, 2005; Dahlander & Wallin, 2006; Tapscott & Williams, 2008).

In licensing transactions, the patent owner keeps the ownership right and transfers only the right to use to a licensee. With this licensing transaction, patent owners can realize complementary value from their patents or multiply the technology. Several value generating options of licensing can be identified. The enforcement of patents is one option to generate financial value from patents. Stick (also known as enforcement or assertion) licensing is based on actual patent infringements and can be conducted by companies that have evidence of a patent's use without authorization (Gassmann & Bader, 2011; Lichtenthaler, 2010; Reinhardt, 2008). The transaction party has already developed and marketed a product. Therefore, stick licensing is a reactive and offensive approach that gives the licensee the right to use the patent without any additional technology or knowledge transfer (Gassmann & Bader, 2011).

Besides generating financial value, the original patent owner can also generate strategic value with patent licensing. In contrast to the reactive and offensive stick licensing, a company can choose the option of carrot licensing, also known as opportunity or enablement licensing (Gassmann & Bader, 2011; Lichtenthaler, 2010; Reinhardt, 2008). Carrot licensing is an active and defensive approach and the original patent owner searches for potential licensees interested in the technology (Gassmann & Bader, 2011). The potential licensee does not use the patent before the licensing contract is concluded. According to Anand and Khanna (2000), licensing is the most significant method for technology transfer and the most commonly observed contractual agreement between companies. The strategic value for the patent owner is generated through the enabler function of the carrot license. The licensee, a company potentially in other industries or markets, is able to commercialize a new product based on the license. Therefore, the original patent owner can enter new markets or applications without producing the product (e.g., Adam, Ong, & Pearson, 1988; Contractor, 1980). Often the licensee has more capabilities in the specific markets or the specific applications (Gredel et al., in press).

In addition to internal exploitation, a patent owner can use these patents to establish an industry standard based on their technology (e.g., Blind & Thumm, 2004; Conner, 1995; Reitzig, 2004a). Even if the technology is already commercialized in products, in several industries, a successful penetration of the market is only possible if other companies also use this technology (Ehrhardt, 2004). With licensing agreements between the relevant players and any third party, it is possible to spread the technology, create a standard, and ensure that consumers adopt the technology.

Patents can also be the basis for co-operation between companies, for example, in joint ventures. Instead of injecting liquid assets into the joint ventures, patents can be used as initial currency (e.g., Parr & Sullivan, 1996; Reinhardt, 2008). Patents exploited in joint ventures can generate strategic value. A joint venture is often set up to develop or commercialize new products and strengthen the market position of the involved companies (Granstrand, 2000). Additionally, spillover R&D effects may occur and the results, or partial results, from the joint venture transferred back to the original patent owner and fuel the internal R&D (Koruna, 2004).

Besides the financial and strategic value, a company can also generate defensive value in licensing transactions (Lichtenthaler, 2011). In complex industries, many products include a large range of technologies. A single firm is no longer able to develop all needed technologies internally and in cumulative technology fields, the innovations build on another (Grindley & Teece, 1997). All companies are dependent on each other (e.g., Cohen et al., 2000; Kash & Kingston, 2001). To generate freedom to operate, prevent infringement and thus enforcement lawsuits, a company can exploit their patents, besides internal exploitation, in cross-licensing agreements (e.g., Grindley & Teece, 1997; Hall & Ziedonis, 2001; Reitzig, 2004b; Shapiro, 2001).

Figure 5 summarizes the different value generating options companies can chose to leverage their patent portfolios. The options are organized regarding the value this choice creates and the strategic need that the patent owner has.

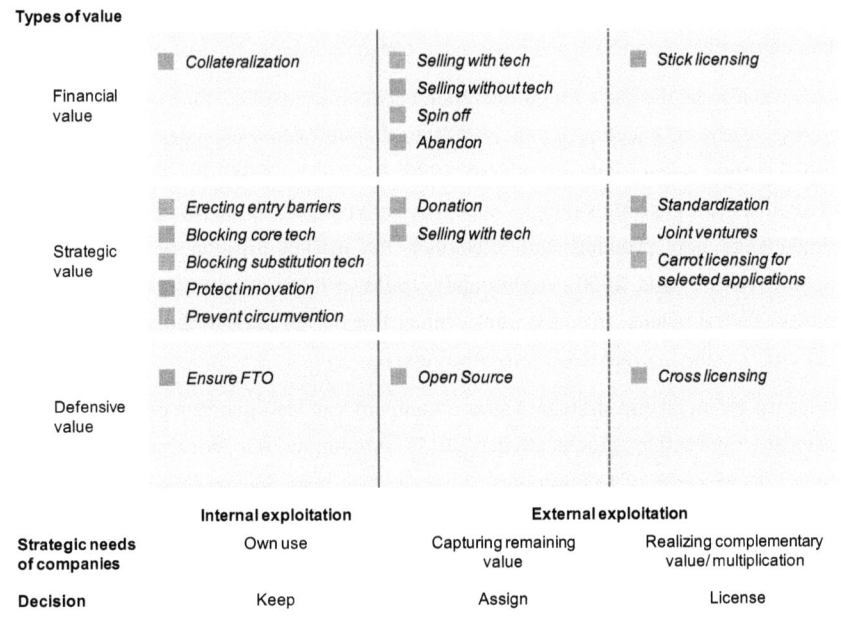

Figure 5: Map of value generating options for leveraging patent portfolios

2.2.2 Impediments to optimally leverage patent portfolios

Even though firms are aware of the increasing importance of patents and the potentials of external patent exploitation, several authors assume that companies do not use all their patents in the patent portfolio optimally and still have large leveraging potentials (Rivette & Kline, 2000). Giuri et al. (2007) analyze the value and use of more than 9,000 European patents and find that on average, 17.4% of all patents in a patent portfolio are exploited neither internally in products or for blocking competitors, nor for licensing or cross-licensing. Whereas in the mechanical engineering industry, the portion of unused patents is 14.3%, patent portfolios in the pharmaceutical and chemical industry have 22.3% unused patents and an even larger optimization potential (Giuri et al., 2007). That patents are not leveraged optimally is often not a function of ignorance but a function of incapability. Berneman et al. (2009) find that in general, companies are willing to exploit 40% of their patent portfolio (on average) externally, but are mostly not able to conduct the exploitation of their patents on their own. The reasons why so many companies have problems acting in the market for patents and technologies are diverse and range from the patent as economic good itself to inefficiencies in the market.

Patents are inherently creative, unique, and idiosyncratic in their scope, depth, strength, and importance (Sneed & Johnson, 2009). These characteristics distinguish patents substantially from tangible assets and cause the difficulties for firms to exploit patents externally. Early studies find that because of the uniqueness and idiosyncrasy of patents, companies have major difficulties in identifying potential marketable technologies (e.g., Arora et al., 2001a; Ford & Ryan, 1977; Lichtenthaler, 2007a; Teece, 1998). Ford and Ryan (1977) also find that companies are not able to do a successful marketing of intangible products.

One significant attribute of patents is that they are highly context specific (Arora et al., 2001a). Therefore, the utilization potential and the value of a patent can vary according to the respective situation, the respective company environment, and the respective patent owner (Arora et al., 2001a; Reitzig, 2003). in transferring patents, technologies, and knowledge to a different context, a resource intensive adaption to this context may be required. in addition, many patents are not suitable for a transfer to a different context and are only relevant for the application they were applied for (Rings, 2000).

Another obstacle that prevents external patent exploitation is the existence of uncertainty. Patent transactions are hindered through uncertainty regarding a patent's value and its tradability (Troy & Werle, 2008). Legal and technical validity determine the intrinsic value of a patent, and they can be both subject to change (Jarboe & Furrow, 2008). Even after granting a patent, this decision can be reversed at a later point in time through oppositions of competitors, incomplete documentation, or plagiarism. Therefore, the legal validity is not guaranteed during the life of a patent, a fact that leads to uncertainty (Jarboe & Furrow, 2008). Technical validity refers to a patent's position within existing knowledge. Patents are granted if they are initially new and not part of the body of knowledge. As technology changes, improves, and moves on, a patent may be no longer new and become obsolete when the underlying knowledge is replaced by more advanced innovation, leading to uncertainty (Arora & Gambardella, 2010a).

Patent valuation assigns a monetary amount to the patent that reflects the economic value of the patent (Ensthaler & Strübbe, 2006). Therefore, even though a patent is protected legally and up-to-date technologically, the economic and strategic value of a patent is still subject to uncertainty (Troy & Werle, 2008). Valid patents do not necessarily generate cash flows, which are often used to calculate the monetary value of patents, and the general ability to generate cash flows is difficult to predict. This is especially significant in industries with a complex technological environment and fast-changing consumer tastes (Jarboe & Furrow, 2008). Hence, the uncertainty regarding future cash flows is large.

In general, patent valuation and the determination of transaction prices are pre-conditions for patent transactions (Kamiyama et al., 2006). However, a standard valuation approach and reliable data on past transactions are lacking (Arora et al., 2001a). Literature offers several patent valuation methods based on the three basic approaches, cost, market, and income approach (for further information see for example, Parr & Smith, 2008), but all methods cannot overcome the general problems of how to deal with the uniqueness of patents (Granstrand, 2000; Troy & Werle, 2008) or the uncertainty regarding future economic benefit (Arora et al., 2001a; Pitkethly, 1997). The absence of valuation methods results in the situation where the value of a patent is not determinable in an objective way (Pitkethly, 1997).

Additional to the firm's internal problems and the contextual factor of patents, structural factors of the market, often intertwined with the first two, arise. Due to the complexity of technologies and inventions, as well as their own characteristics, patents are difficult to evaluate for companies and persons that have not participated in the development of the invention. Hence, the economic value of a patent is difficult to estimate for outsiders and asymmetric information between patents buyers and patent sellers exist (Caves et al., 1983; Troy & Werle, 2008). As a result, information asymmetries prevent efficient market clearing (Tietze, 2011). In general, markets for patents and technologies lack transparency regarding essential market information. Companies willing to trade are not able to gather information about buyers, suppliers, and technologies and patents offered (Lichtenthaler & Ernst, 2008a).

In summary, uncertainty, problems in patent valuation, and the absence of data on past transactions lead to a lack of transparency in essential market information and efficiency in the market for patents and technologies and to high transaction costs for the actors in the market for patents and technologies (e.g., Arora et al., 2001a; Arora

& Gambardella, 2010a; Caves et al., 1983; Ford & Ryan, 1981; Gambardella, 2002; Lichtenthaler & Ernst, 2007; Monk, 2009; Tietze, 2011; Troy & Werle, 2008).

2.3 Third parties as enablers of transactions

Along with the development of the market for patents and technologies, its high transaction costs, and its lack of transparency, a new business model has emerged: technology market intermediaries (e.g., Hargadon & Sutton, 1997; Howells, 2006; Nambisan & Sawhney, 2007; Sapsed et al., 2007). In general, technology market intermediaries have a broad focus and transfer technology, innovation, and patents equally (Benassi & Di Minin, 2009; Monk, 2009; Wang, 2010). According to Chesbrough (2006), these companies are called innovation intermediaries. Their main function is to support owners of a technology to find a buyer or licensee. Patents are the legal mechanism for transferring technologies (e.g., Arora, 1995; Chesbrough, 2003b), and a sub-group of these intermediaries focuses more on the transfer of patents.

Technology market intermediaries focus on business-to-business related transactions and facilitate the transactions of patents (Tietze, 2011). Therefore, companies can utilize them to leverage optimally their patent portfolios (e.g., Benassi & Di Minin, 2009; Kelley, 2011; Millien & Laurie, 2008; Yanagisawa & Guellec, 2009). The development and role of technology market intermediaries is investigated from different perspectives and a number of different sources. Five major fields of research that analyze the role of technology market intermediaries can be identified.

Literature on technology transfer and diffusion focuses on the influence of intermediaries on the speed of diffusion and the reception of new products (Hägerstrand, 1952), as well as the complementary skills an intermediary can offer in technology transfer processes (Shohert & Prevezer, 1996). Literature on innovation management focuses on the innovation process and on which activities intermediaries are involved in. In addition, the role of intermediaries as facilitators of the knowledge transfer process is emphasized (Hargadon & Sutton, 1997). Literature on systems and networks of innovation focuses on the economic impact of intermediaries and on how intermediaries influence the entire innovation system. Intermediaries support the information flow, and they are linked with principal agent models (Klerkx & Leeuwis, 2009; Lynn, Mohan Reddy, & Aram, 1996; Stankiewicz, 1995). Literature on service organizations focuses on the role of intermediaries in the context of service innovation and service activities (Bessant & Rush, 1995). Literature on patent litigation focuses on the role of intermediaries as creators of credible threats of litigation and providers of liquidity in the market for patents. This stream of literature focuses on intermediaries, which transfer the sole patent right without additional know-how or technology (McDonough III, 2006; Shrestha, 2010). The following part aligns the different streams of literature and describes the business models of technology market intermediaries and non-practicing entities, a controversy discussed sub-group of patent intermediaries.

2.3.1 Bridging patent supply and patent demand

Technology market intermediaries are agents that fulfill a wide variety of tasks and functions in the external patent exploitation process between two or more partners (according to Howells, 2006). The literature is highly fragmented, and a large variety of terms for this type of agent exists. They are also called intermediary firms (Stankiewicz, 1995), bridgers (Bessant & Rush, 1995), brokers (Benassi & Di Minin, 2009; Hargadon & Sutton, 1997), or superstructure organizations (Lynn et al., 1996). In the literature, technology market intermediaries are mainly associated with innovation processes or external technology exploitation projects.

In general, technology market intermediaries are defined as organizations that match the supply and demand of patents, in combination with or without technology or additional knowledge. Therefore, they aim to facilitate patent based transactions. These organizations do not innovate, develop technologies, or conduct contract research (according to Benassi & Di Minin, 2009). The activity of patent intermediation is the core activity of technology market intermediaries, not only a corresponding service (Winch & Courtney, 2007). Their position in patent transfers is distinct (see Figure 6).

Third parties as enablers of transactions 39

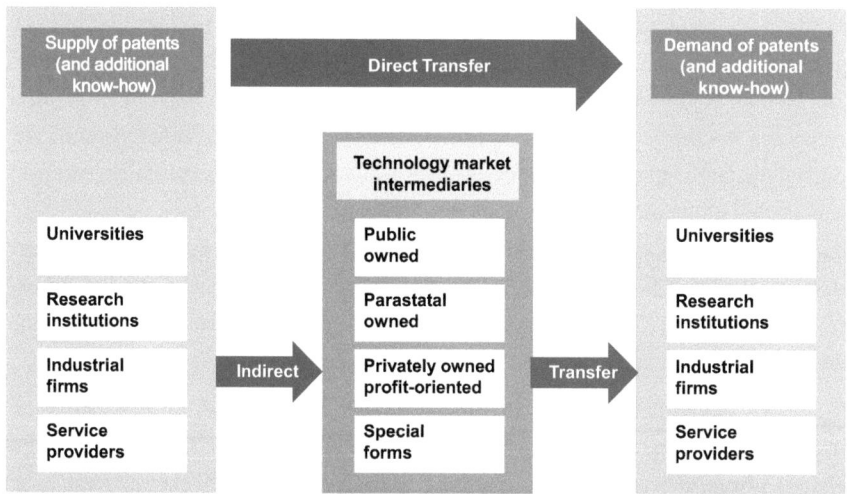

Source: According to (Reinhard & Schmalholz, 1996)

Figure 6: Transfer of patents and technology market intermediaries

The direct transfer of patents indicates a transaction between two parties without the participation of a third party as a middleman. Transactions of patents conducted with an intermediary between two parties are denoted as indirect transfer of patents (see Figure 6).

Due to high transaction costs and the resulting imperfections in the market for patents and technologies, the concept of intermediaries has been transferred from financial markets, where it first occurred (Stigler, 1951), to the market for patents and technologies. External patent exploitation is more complex than commercializing goods on product markets (Callon & Muniesa, 2005). To conduct a transaction successfully, firms can develop their own competencies or rely on the services of technology market intermediaries (Nambisan & Sawhney, 2007; Sapsed et al., 2007). Intermediaries have accumulated experiences in the market for technology (Morgan & Crawford, 1996); therefore, they may contribute to reduce the inefficiencies in the market for technology (Bryant & Reenstra-Bryant, 1998). From a transaction costs economics perspective (Williamson, 1975), technology market intermediaries could facilitate the markets by reducing operative costs benefiting from economies of scale and scope and bargaining asymmetry (Benassi & Di Minin, 2009).

The functions performed by technology market intermediaries depend on the type of agent, the type of business model, and if patents are transferred with or without know-how. Morgan & Crawford (1996), state: "Technology broking is not a well-defined activity and the heterogeneous nature of the participants is a key characteristic of the industry" (p. 363). Some intermediaries offer services supporting the entire external patent exploitation process, others serve as additional resources for specific tasks (e.g., valuation of patents and determination of transaction price). For the innovation process, Howells (2006) identifies ten functions of technology market intermediaries. These ten functions can be clustered into three categories: (1) facilitating collaboration, (2) connecting, and (3) providing service (Lopez-Vega, 2009).

Only a few publications attempt to systemize technology market intermediaries or patent intermediaries. Benassi and Di Minin (2009) conceptualize the heterogeneous activities of patent intermediaries according to the characteristics 'commitment' and 'value added to the exploitation process' and derive seven different types of patent brokers. Those seven patent intermediaries include two business models that aggregate patents: aggregators and the enforcers. This study shows also that most authors allude to the topic of patent aggregating companies in their investigation of patent intermediaries (e.g., Gredel et al., in press; Wang, 2010; Yanagisawa & Guellec, 2009).

Kelley (2011) identifies patent aggregating companies in her analysis of the player in the patent marketplace. She categorizes the players into buyers, sellers, and facilitators. Patent buyers are either financial buyers or other buyers. According to their motivation, Kelley (2011) distinguishes financial buyers roughly between patent assertion firms, defensive aggregators, and *Intellectual Ventures* without analyzing activities or motives in detail.

Analyzing the activities of patent brokers, Benassi and Di Minin (2009) identify several activities conducted by patent intermediaries, such as patent valuation and evaluation, selection of patents, negotiation with transaction partners, or assisting transactions. With regard to patent aggregating companies, the authors find that their activities exceed the completion of contracts as the patent aggregator and the patent enforcer takes over high risk. Among the 19 business models Yanagisawa and Guellec (2009) identify, three business models focus on aggregating patents: patent pool administration, IP aggregation and licensing, and defensive patent aggregating funds and alliances. However, Yanagisawa and Guellec (2009) describe the activities of these business models only shortly without depth and empirical data. To discuss the role patent intermediaries play in the patent market, Wang (2010) splits patent intermediaries into companies that assist patent acquisitions (brokers) and those companies that acquire patents. The latter group is further divided into defensive aggregators, companies that acquire patents to provide their subscribers with freedom to operate, and offensive aggregators, companies that develop and acquire patents to realize revenues through licensing and asserting patents. In addition, Wang (2010) focuses mainly on identifying the players and does not give information on the activities of technology market intermediaries or patent aggregating companies. Gredel et al. (in press) analyze patent-based investment funds and their position as innovation intermediaries for SME. Analyzing two case studies, they identify the type of targeted patents. Motives to cooperate are analyzed from the position of the original patent owner that aims to improve their financial situation. The analysis of activities of the patent-based investment fund is set to the advancement of the acquired technology.

2.3.2 Non-practicing entities and their intermediation of patent transactions

During the last decade, a business model of patent intermediaries has developed that has provoked controversy. The points of contention range from ethical issues regarding the use of the patent system, to definitional issues if these companies are patent intermediaries at all. The companies in question do not produce but they exploit patents externally. According to their external exploitation strategy, non-practicing entities enforce patents vigorously. Their actions are legal and within the system. The main point of criticism on these companies is that they do not try to exploit the invention itself and use the patent in its historical meaning hence, to protect the innovation from imitation. In the literature, authors use the terms non-practicing entities (NPE) (e.g., Shrestha, 2010), non-producing entities (Johnson et al., 2007), patent trolls (e.g., Fischer & Henkel, 2009), or patent sharks (e.g., Reitzig et al., 2007). However, the range of business models of these companies is wide, and a common definition does not exist. The different terms used for these companies are often negatively connoted.

Central to the description of a non-practicing company is that they use patents that do not cover their own products to gain revenues from licensing. Hence, they do not practice their patents. Some companies create these patents in their own R&D departments (e.g., *Rambus, Qualcomm, Tessera*); others buy patents from distressed securities or other willing-to-sell firms and enforce these patents (e.g., *IPCom, Millennium IP, Ronald A Katz Technology Licensing*). Between these two types, a wide field of business models exists.

Definitions are heterogeneous, and several authors attempt to name the company and to describe the business model. Former Vice President and Assistant General Counsel of Intel Peter Detkin coined the term 'patent troll', which is widely used today. Even though the term 'troll' is heavily connoted, it is used in academic literature (e.g., Chien, 2009; Fischer & Henkel, 2009; Geradin et al., 2011; Golden, 2007; Gregory, 2007; Lemley, 2007; McDonough III, 2006; Merges, 2009; Reitzig et al., 2007; Shrestha, 2010). For example, Reitzig et al. (2007c) describe patent trolls as "patent holding individuals or (often small) firms who trap R&D intensive manufacturers in patent infringement situations in order to receive damage awards for the illegitimate use of their technology" (p. 134).

> [....] We denote patent sharks or trolls as individuals or firms that seek to generate profits mainly or exclusively from licensing or selling their (often simplistic) patented technology to a manufacturing firm that, at the point in time when fees are claimed, already infringes on the shark's patent and is therefore under particular pressure to reach an agreement with the shark. (Reitzig et al., 2007, p. 137)

The term non-practicing entity is more neutral than patent troll is. However, the literature is ambiguous whether a NPE is the same type of company as a patent troll. Magliocca (2007) states that a NPE is a troll: "There is simply no way to subdivide NPE into 'good NPE and bad NPE'. There is no judicially-manageable bright line between supposed 'patent trolls' and inventors who cannot practice their inventions because of resource limitations or managerial considerations" (p. 52). Other authors differentiate between NPE and troll. For instance Layne-Farrar and Schmidt (2009) state:

> This result leads us to reject the prevalent definition of a patent troll as any non-practicing or non-innovating entity. Indeed, NPE are the least likely to exhibit troll behaviors. Instead, a better gauge is the presence of special conditions for a patent hold-up and the exploitation of irreversible investments, regardless of the business model of the patent holder. (p. 1139)

From an extensive literature review, the following criteria, used to distinguish between trolls and non-trolls, are identified:

- *Non-producing* (e.g., Bessen, Ford, & Meurer, 2011) – A company is a patent troll if it gains revenue from licensing but does not produce goods or services. That leads to difficulties with the classification of single inventors, think tanks, and universities.
- *Products are not commercialized* (e.g., Golden, 2007) – This criterion seems to be similar to non-producing but many firms do not produce their products themselves. They are only orchestrators that have out-sourced their production and do only selected steps in the value chain (e.g., Nike, Adidas, Apple).
- *Does not conduct own R&D* (e.g., Rubin, 2007) – A company is a patent troll if it gains revenue from licensing but does not have own R&D. This criterion excludes think tanks, universities, and inventors clearly from being a troll.
- *Acquires patents* (e.g., Fischer & Henkel, 2009) – A company is a patent troll if it buys patents instead of own development. However, this includes many producing industry firms because buying patents is central for open innovation.
- *Owns patents which do not cover the core business* (e.g., Rubin, 2007) – This criterion could include companies that have more business divisions or small companies that are dependent on each single invention.

Additionally, literature defines trolls by analyzing their behavioral patterns:

- A troll does not target ex ante license revenues but searches for infringements (Reitzig et al., 2007). This increases the revenues because firms may not have an alternative to pay, otherwise they would have to shut down production.
- A troll litigates low quality patents and gains overly licensing revenues (Fischer & Henkel, 2009).

Academic literature focuses either mainly on patent characteristics that are interesting for NPE, or on the business models of NPE. Describing the business model, NPE can be seen as opportunistic licensers that are able to benefit from a large gap between the acquisition price of a patent and the royalties it receives in patent enforcement cases (Magliocca, 2007). Thus, NPE may simply be seen as corporations that acquire undervalued patents in an attempt to profit through licensing and enforcement (Johnson et al., 2007). According to Magliocca (2007), companies have this arbitrage opportunity if the costs of patents are low, if substituting the disputed technology is unreasonable, and the outcome of infringement litigation is uncertain. Even though these characteristics are quite general, it is important to state that companies are involved in this arbitrage scheme if the profitability is large and higher revenues offset the costs. Therefore, NPE target industries and technological areas with comparable high revenues.

Literature distinguishes a three-step process for the patent aggregating activities of NPE: building up patent portfolios, waiting for producing companies to infringe the patents, and enforcing the infringed patents (Henkel & Reitzig, 2008). In the first phase, NPE build up large patent portfolios either from patenting their inventions or from acquiring or exclusively in-licensing patents from other corporations. In the second phase, NPE either screen the market to detect already occurred infringements or wait until infringement occursHenkel and Reitzig (2007) find that NPE advisedly wait until infringing companies face the already high costs of substituting the infringing technology. As the costs of inventing around or substituting unreasonable high, the infringing company would rather take a license. As Henkel and Reitzig (2007) state, "when patents are hidden, companies unknowingly lack vital information when creating new products", (p. 131), which leads to inadvertent infringement. Therefore, critics accuse NPE of timing their invention, waiting until the infringed product is on the market, and of when costs of abandoning the product are too high. At this point, the infringing company is under high pressure, and the likelihood that it takes a license is high. In the third phase, NPE contact or immediately sue the infringing company and offer a license or a settlement. In infringement cases between producing companies, the defendant and plaintiff often agree on cross-licensing. Due to the fact that NPE do not need access to the other parties' patent portfolios to produce, the infringing company either has to settle or to litigate (Luman III & Dodson, 2006). NPE are in a good position because condemnation payments, the threat of injunction, the high litigation costs, and the uncertainty regarding the outcome of the lawsuit drive companies to settle the case and take a license.

Part of the controversy discussion on NPE is that there is disagreement amongst scholars regarding the benefit and detriment of NPE. From an original patent owners' perspective, proponents argue that NPE provide capital and bargaining power to single inventors and that SMEs that lack the resources to enforce patents themselves (Ball & Kesan, 2009). Enforcing the patents fosters innovation and technological progress and rewards single inventors and SMEs with returns on innovation. NPE transfer patents and licenses from the original patent owner to companies that already use the patents, and therefore, conduct forced intermediation. Based on the force factor, from a patent user perspective, opponents argue that NPE distract producing companies from their core business (Williams & Gardner, 2006). As patent enforcement is the core business of NPE, they can focus revenue generation from it, whereas producing companies have to shift resources no longer available for core activities. Additionally, producing companies face the threat of permanent injunction, which could lead to unreasonable actions on the part of the producing company (Luman III & Dodson, 2006). In sum, this behavior leads to higher product prices because producing companies often pass the costs of royalty payments and patent litigation to consumers (Davis, 2008).

Also on a macro level, the benefits or detriments of NPE are discussed. Advocates draw on the argument that NPE aggregate patents. Hence, NPE create a market for patents. With an increasing number of transactions, patent valuation methods may improve and transaction costs decrease. Thus, NPE foster the development of the market for patents and technologies (Fischer & Henkel, 2009; McDonough III, 2006).

Even though patents are transferred in the market for patents and technologies, opponents invoke that NPE acquire and litigate only weak or obscure patents with a broad scope (Shrestha, 2010) therefore, harming the innovation system. Patents with a broad scope have a higher likelihood that a larger number of products and processes will infringe upon it (Merges & Nelson, 1990). Fischer and Henkel (2009) show that NPE prefer to acquire patents with a broad scope. However, compared with litigated patents from producing companies, NPE's patents are, on average, of higher technological quality and legal sustainability.

2.4 Reference framework

This research aims at contributing to literature and theory on patent management, markets for patents and technology, and as technology market intermediaries. As such, it tries to answer the question whether patent aggregating companies can be utilized for patent portfolio leveraging activities of producing companies. Therefore, the phenomenon of patent aggregating companies has to be explored, and information on strategic, organizational, and operational aspects has to be gathered. In order to produce sound results, as well as to facilitate and guide data collection and analysis, a reference framework is constructed (Miles & Huberman, 2004). In addition, it enhances the understanding of the phenomenon and allows for a broader evaluation of the relevant aspects within the empirical data. Therefore, the framework is based on insights from a broad literature review and it thus builds the foundation for data collection.

The reference framework is based on the literature on patent management and on technology market intermediaries (see Figure 7). Literature on patent management serves as basis for the context between the producing company and the patent aggregating company. Additionally, this stream of literature emphasizes the importance of the motives and strategies regarding patenting and the acquisition of patents for companies. The literature on technology market intermediaries implies that operational aspects of intermediaries are the major factor for the producing companies' utilization decisions. Operational aspects, in combination with organizational aspects, are the key determinant for the success of patent aggregating companies.

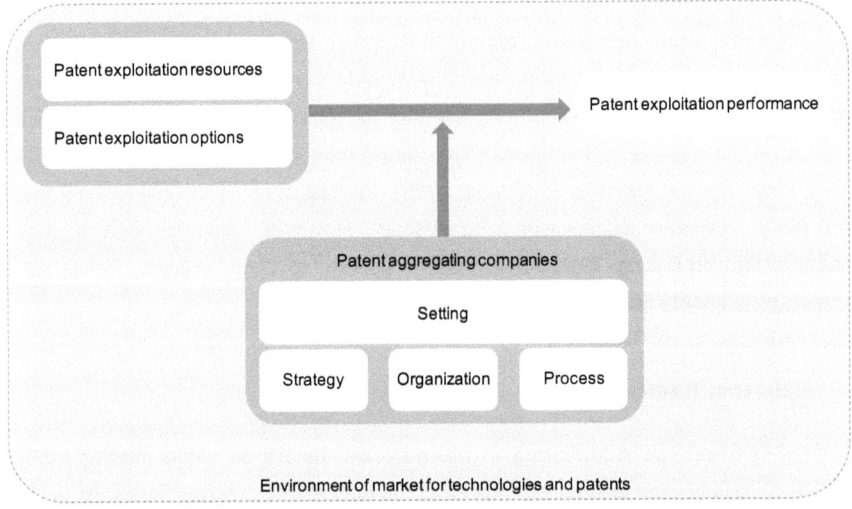

Figure 7: Reference framework to analyze patent aggregating companies

The reference framework serves as the basis for the empirical investigation in the case of firms and serves as a guideline to ensure that all topics relevant for the exploration of patent aggregating companies are covered. According to the nature of explorative research, the process of data gathering and data analysis is iterative. Therefore, the reference framework is adopted on the basis of new information resulting from the analysis of the empirical data. Adapting the reference framework is based on the aim to reflect reality better and serves as a sound base to develop results that extend the extant literature and theory on patent management and technology market intermediaries.

3 Exploring the phenomenon of patent aggregating companies

As patent aggregating companies have emerged as a recent empirical phenomenon, the knowledge of their activities and strategies is limited. These companies are still a black box; therefore, their utilization by producing companies is difficult. To shed light on patent aggregating companies, this chapter provides a comprehensive analysis of patent aggregating companies and their settings, activities, and strategies. Based on empirical data from interviews and second sources, the first part of this chapter illustrates a general picture of patent aggregating companies. The second part describes the process of patent aggregation. The last part illustrates the strategies of patent aggregating companies and derives eight business models.

Appendix 1 gives an overview of the patent aggregating companies that were analyzed to detect general patterns, the setting the companies operate in, their processes and activities, and their strategies.

3.1 Setting of patent aggregating companies

Before exploring the activities and strategies of patent aggregating companies, the following part provides general information about the age, history, and geographical location of the sample companies. Light is also shed on the different ways of venture creation and funding of the analyzed patent aggregating companies.

3.1.1 General information

The dates of formation of the analyzed companies support that patent aggregating companies are a young phenomenon: 20 out of 27 patent aggregating companies were founded in or after 2000. Figure 8 provides an overview on the number of founded patent aggregating companies per year.

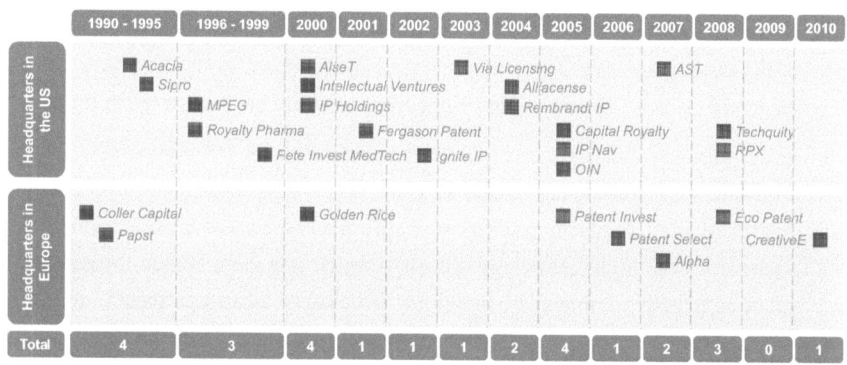

Figure 8: Year of formation and geographic location of sample companies

Analyzing the history of the seven companies founded before 2000 more closely, only *Sipro Lab*, *Pete Invest MedTech*, and *MPEG LA* were in the patent aggregating business before 2000. At date of formation in 1993, *Acacia Research* was a venture capital firm and started today's business model in 2003. *Coller Capital* was founded as a private equity firm in 1990. It became involved in patent aggregation in 2006 when an executive with experience in leveraging research and technology assets joined the management team. *Papst Licensing* started to monetize the patents of the manufacturing mother company in 1993 and became involved in patent aggregation in 2000. Founded in 1996, *Royalty Pharma* closed its first patent deal in 2000.

Figure 8 shows two peaks of funding activities: one in 2000 and one in 2005. The turn of the millennium is also known as new economy or dotcom era. This time is characterized by the rise of startup companies from so-called sunrise industries as information technology, multimedia, biotechnology, or telecommunication. A high research intensity and large numbers of patent applications characterize these industries. When the dotcom bubble burst in 2000, many of the new economy ventures went insolvent and patents were available for little money. After a worldwide recession, the economy recovered in 2004 and 2005. In this time, R&D expenditures and patent applications revived and investors returned to the markets. These two general economic conditions could have fostered the formation of patent aggregating companies.

Most of the analyzed patent aggregating companies are headquartered in North America. Figure 8 shows that 18 companies are headquartered in the US, only *Sipro Lab* is located in Canada. The remaining eight companies are headquartered in Europe.

Additionally, regional clusters can be observed. From the 18 US companies, 7 are located in California, 6 on the east coast, and 3 in Texas. This distribution reflects the characteristics of the US patent ecosystem. California, especially Silicon Valley, is home to the American high-technology industry, which is a large source for patents. Many patent aggregating companies, therefore, are closely located to their source of patents. During the last years, Wall Street has recognized patents as a financial asset (Yurkerwich, 2008). Therefore, New York is an interesting location for business models in these areas. The third cluster located in Texas can be explained by the US court system. The US District Court for the Eastern District of Texas has become known for patent litigation lawsuits (Barry et al., 2010; Taylor, 2007; Williams, 2006) and therefore, offering new business opportunities for companies focusing on patent assertion.

All companies gave, at least rough, information on their number of employees but only a few of the analyzed companies agreed on disclosing quantitative information regarding their operation (12 out of 27). The average size of the companies is around 20 employees. Whereas *Intellectual Ventures* has the most employees (ca. 800), most other firms (23 out of 27) have between 5 and 50 employees. The median is ten employees and supports the observation that patent aggregating companies operate with few employees. The number of acquired patents or patent portfolios is confidential for most of the companies. *Intellectual Ventures* seems to be the biggest player in the patent market. In private transactions, but also as main buyer at the *Ocean Tomo* Live Patent Auctions (Ewing, 2010), *Intellectual Ventures* has acquired a patent portfolio of several thousand patents. In July 2011, *Intellectual Ventures* announced that its patent portfolio consists of more than 35,000 US and international patents and patent applications. *Intellectual Ventures* is also the biggest spender in the sample and seems to be the biggest spender in the market for patents (Benassi & Di Minin, 2009; Holden, 2011; PatentFreedom, 2011b; Yurkerwich, 2008). Until now, the company has spent ca. EUR 1.15 billion to acquire patents and patent applications.

3.1.2 Venture creation and funding of patent aggregating companies

Different paths of venture creation can be observed (see Figure 9 for a graphical overview of the different paths). Only 2 out of the 27 patent aggregating companies are continuations of pre-existing activities. As mentioned above, *Acacia* was founded as a venture capital company with the focus on dotcom companies. After the burst of the technology bubble, many investment companies failed and *Acacia* was stranded with the patents. Out of this situation, *Acacia* changed its business strategy and no longer invested in R&D and start ups but started to out-license the stranded patents. Being successful with this business model, *Acacia* acquired more patents and became a patent aggregating company. Acacia realized

> there was a huge market need for an 'outsourced patent licensing' company to assist patent owners in generating licensing revenues from their patented technologies. From that point forward we focused our companies' efforts exclusively on building the 'leading outsourced patent licensing company'. (Ryan, 2011).

The other company that changed its business model is *Papst Licensing*. *Papst Licensing* evolved from the producer of small electric motors and electronic cooling fans, *Papst-Motoren GmbH & Co KG*. In the beginning of the 1990s, this producing company faced financial difficulties and had to sell its producing business to a competitor (*EBM*). Around 600 patents and pending applications were acquired from *Papst Motoren* and transferred to the newly founded company *Papst Licensing*, which was founded to monetize and enforce these patents through out-licensing. After gaining experiences in patent licensing and enforcement, *Papst Licensing* started to offer its services to third parties and to acquire their patents.

In 4 out of the 27 patent aggregating companies, amassing patents is an additional business model to the original existing business model. *Coller Capital* is a private equity business, but with an overall increasing interest in IP and a new partner with patent market experience, the IP investment group was established. In addition, the patent aggregating business model of *Pete Invest* was set up as an additional product.

The remaining 21 patent aggregating companies are founded with their recent strategies and business model. The creation of these patent aggregating companies can be divided into three paths of venture creation. Professionals or companies already working in patent transactions pursued the first path of venture creation. By founding a patent aggregating company, they leveraged their experiences and knowledge and additionally diversified their existing businesses. This is, for instance, the case of *Alliacense*, a fully owned subsidiary of the *TPL Group*. The *TPL Group* is a service provider for IP management founded in 1988. Through *Alliacense*, the *TPL Group* now manages the licensing programs of five patent portfolios containing around 120 patent families.

A group of companies created by daring entrepreneurial activities and often backed by large financial resources pursed the second path, and these can further be divided into two subgroups. On the one hand, professionals, often attached to multinational technology corporations, with deep insides in the patent market spotted the emerging opportunity of the patent aggregating business, drafted a business plan, and were able to attract large amounts of funding. For instance, the founders of *Intellectual Ventures* used to be senior managers of *Microsoft* and *Intel*. Based on their business plan, their knowledge, and their contacts, they were able to collect large funds of corporate investors, such as *Apple*, *Nokia*, *Sony*, and *Microsoft*, as well as financial investors, such as *JP Morgan* or *Charles River Venture*. On the other hand, entrepreneurs or financial institutions that recognized patents as an asset class ahead of time have founded patent aggregating companies. Their path of venture creation was firstly to close an investment fund with institutional investors or high net worth individuals and secondly, to acquire large amounts of patents. *Patent Invest Fond* is an example for the later subgroup. Finance System initiated the funds of Patent Invest Fond. Sales partner *Credit Suisse* collected ca. EUR 20 million in total.

The third group includes companies or individuals that teamed up with other companies or individuals to pursue the same goal. Based on increasing patent enforcement activities, high-technology companies often face the threat of difficult licensing negotiations, permanent injunctions, or expensive litigation lawsuits. To reduce these threats, several high-technology companies, among them *Ericsson*, *IBM*, *Hewlett-Packard*, *Motorola*, and *Sun Microsystems,* founded *Allied Security Trust*. *Allied Security Trust* is a member based defensive organization that acquires patents that could be a threat to its members.

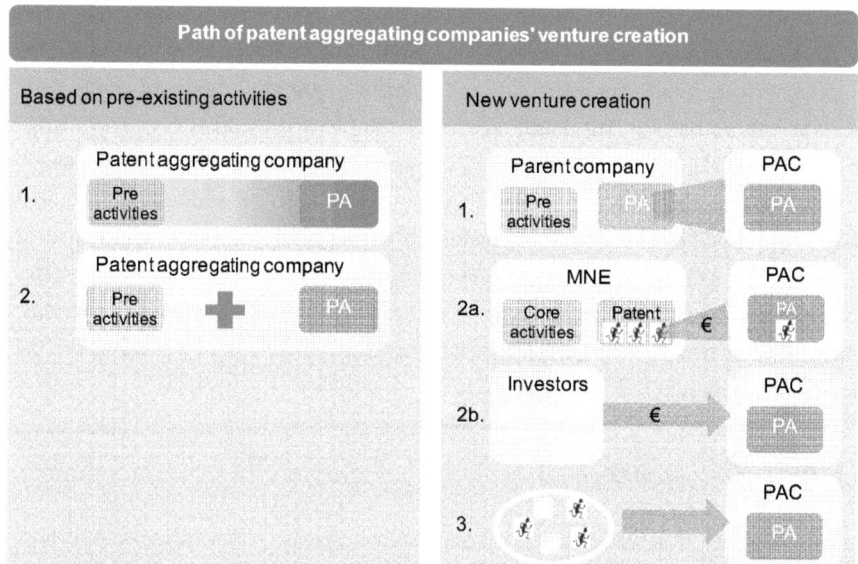

PA - patent aggregating activities
PAC - patent aggregating company

Figure 9: The different paths of venture creations

Even though the acquisition of patents requires significant financial resources, different models of funding can be observed. In addition, the ownership of the analyzed patent aggregating companies differs. Only 2 out of the 27 companies are public companies. While *Acacia* went public in 2003, the *RPX* initial public offering (IPO) took place recently. On May 5, 2011, *RPX* raised approximately USD 159.6 million, with shares trading at USD 19. Both companies use the generated money for patent acquisition purposes. Before the IPO, the private equity firms *Kleiner Perkins Caufield & Byers* and *Charles River Ventures* funded *RPX*, as well as their patent acquisitions. All other analyzed companies are private entities. Even though they are all privately held, they differ substantially regarding the funding of their patent aggregating activities. In analyzing the funding structures of the patent aggregating activities, five different funding schemes can be identified. The first scheme is the *member model*. For instance, its members founded *Allied Security Trust*. These members also fund the patent acquisition activities. The second scheme is the *capital market model*. The privately held *Capital Royalty*, for example, funds its aggregating activities by securitizing the acquired royalty interests at the capital market. The third scheme is the *venture capital model*. This model, for instance, is applied at *Collar Capital*. Private equity company *Collar Capital* has collected a private equity fund used to acquire patents. The fourth scheme is the *privateer model*. *Fergason Patent Property* is privately owned and started its business exploiting the patents of the single inventor and entrepreneur Dr. James Fergason. Until now, all additionally acquired patents are funded based on prior generated licensing revenues of the original patent portfolio. The fifth scheme is the *volunteer model*. *MPEG LA*, for instance, aggregates the patents licensed by patent owners to licensees and is the medium to administer this. Patent owners are highly interested in pooling their patents, thus a funding for the aggregation activities is not necessary.

3.2 Process of patent aggregation

In analyzing the activities of patent aggregating companies, a general process of how patent aggregating companies amass patents and work with them can be derived (see Figure 10). The process of operation can be subdivided into the following four phases:

(1) *Selection* of patents.
(2) *Structuring* one or more patent portfolios by acquiring ownership rights in patents or commercialization rights.
(3) *Value adding*, the phase where the patent aggregating company accomplishes several activities.
(4) *Exploitation* of patents.

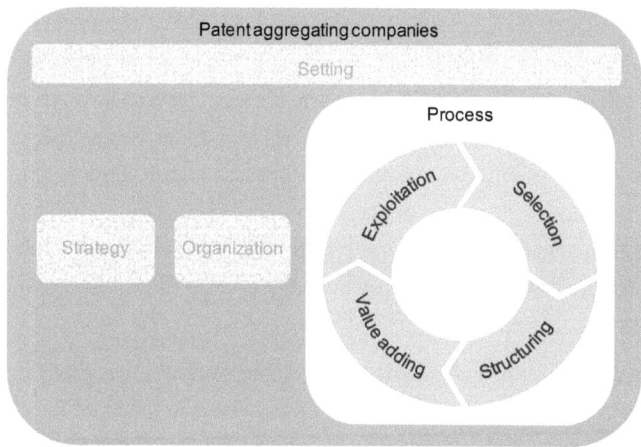

Figure 10: The process of patent aggregation

3.2.1 Selection of patents

In the selection phase of the patent aggregating process, the strategic direction of the planned patent portfolio is shaped. The patent aggregating company determines the industry and the geographical scope, as well as the breadth of transaction of the targeted patents and identifies the patents and original patent owners.

Among the patent aggregating companies, the sectoral focus of the patents they acquire varies. Almost half of the companies in the sample focus on technologies of either the electrical engineering industry, which includes digital media, electronics, information and communication technology, medical devices, semiconductors, or software; or the life science industry, including pharmaceutical products, chemicals, and biotechnology. This specialization reflects the importance of patents in these two industries. In the life science industry, patents are generally used to secure the market power of firms. Firms can gain and enforce important monopolistic status for these industries by using patents. Especially in the pharmaceutical industry, blockbuster products are highly dependent on patents, since patents allow the inventing firm to hinder generic companies in producing the respective drug. High-technology companies have a different use of patents. In electrical engineering, the interdependence between firms resulting from their patent portfolios is very high. Companies are not able to market new products autonomously without being contingent on third party patents. Consequently, companies often apply for patents in order to block competitors, or to strengthen their positions in cross-licensing negotiations. Large numbers of patent applications are filed to prevent inventing around single patents, a practice that results in patent thickets. Patent aggregating companies specialize on patents from these two industries because high-technology or life science patents are very likely to obtain financial returns. The remaining half of the sample's companies also acquires patents from electrical engineering and the life science industry, but interviewees pointed out that they are open to all industries to prevent the exclusion of a valuable opportunity.

In addition to an industry focus, patent aggregating companies focus on specific geographic markets. The US patent market is rich with opportunities to generate large revenues from patent transactions, and patent enforcement seems more lucrative than in Europe. Therefore, most aggregating activities focus on the US market, and patent aggregating companies buy mainly US patents. Many of the patent aggregating companies also have offices in Europe and Asia. These offices deal with local companies to acquire their US patents. Exceptions are *Alpha Patentfonds, CreativE* and *Patent Select*. The patent portfolio of *CreativE* holds mainly patent documents from the UK, Germany, and France. *Alpha Patentfonds* focuses on the acquisition of German patent documents.

The amassed patents can be divided into two categories. Patents in the first category are aggregated to amass the sole legal right of exclusion, without the underlying technology or additional knowledge transfer. Usually in this case, the patent is already granted, and the technology is in a later stage of the technology lifecycle. For instance, *Rembrandt IP Management* is primary interested in acquiring patents that are currently infringed. Based on this approach, *Rembrandt* transfers only the legal rights and does not utilize the technology. Patents in the second category cover technologies in an early stage of the technology lifecycle but also technologies already in use. They are aggregated in combination with the technology and additional knowledge that can be used to develop embryonic technology further. For instance, *Techquity Capital* is primarily interested in patents that demonstrate a fundamental contribution to the technology area they relate to and have the potential to be developed further.

3.2.2 Structuring of patent portfolios

The activities in the structuring phase aim at creating a portfolio that optimally fits the general objective of the patent aggregating company. An optimal structured patent portfolio is one of its success factors. Therefore, this phase focuses on approaching the original patent owner, negotiating and obtaining the patents, and bundling them to powerful patent portfolios.

Most patent aggregating companies approach patent owners actively, but increasingly patent owners become active and offer their patents. Patent aggregating companies are increasingly approached as the name of the patent aggregating company becomes known in public. One example for this shift in the contacting approach is *Acacia*. At the beginning, it was exclusively *Acacia* reaching the patent owners. Based on research of patent attorneys and engineers, the company contacted the owners and started negotiating. Based on its track record and its visibility as a public company, patent owners now increasingly call *Acacia*, and the company has shifted to an passive acquisition approach. Independent from its recognition, *Sipro Lab* follows an active approach. After defining the strategy for the intended patent portfolio, the company publishes 'patent calls' on its website, as well as actively detects and approaches patent owners based on their large network.

Before starting negotiations and bundling activities, the patent aggregating companies evaluate the offered patents extensively. The evaluation criteria and the extent of the evaluation process depend on the strategic direction established for the patent portfolio. In general patent management, patents are evaluated from a legal, an engineering, and a business perspective. For all patent aggregating companies, the legal perspective is of great importance. Without a valid patent, the patent aggregating company would lose the basis for further activities. Depending on the business model of the patent aggregating company, the importance of the technical or the business perspective is ambiguous. *Intellectual Ventures*, for instance, invests in a broad range of industries and evaluates the patents regarding the technical quality of the invention, the legal quality of the patent, the volume and potential of the market a patent is commercialized in, existing litigations, involved parties and results, and the expected performance of a product the patent covers. However, since autumn 2008, the evaluation process contains a showstopper. *Intellectual Ventures* invests only in patents that are commercialized in products, therefore, the criterion 'evidence of use' must be sufficient.

Golden Rice PDP has a different evaluation procedure. *Golden Rice PDP* mainly evaluates if the patents cover the intended technological application. The business potential is only of marginal interest. In contrast, *Patent Select* has a major focus on the commercialization potential of the underlying technology and therefore, relies on criteria like market potential, market volume, quality and potential of invention, or remaining R&D costs, until the technology is ready for the market. The legal dimension of patents is also checked, but the scope of patents is still changeable in further development.

Several patent aggregating companies employ service providers in the structuring phase of the patent aggregating process. Service providers mainly act as middlemen between the supply and demand of the patents, as well as support in the evaluation process. *Allied Security Trust*, for instance, often works with patent brokers. *Allied Security Trust* screens the market with the help of a network of more than 300 brokers that offer patents for sale. The advantage of this proceeding is twofold: on the one hand, *Allied Security Trust* preserves the anonymity of its members. Patent owners often increase their asking price if they realize that a company with substantial financial resources shows interest in their patents. By engaging a patent broker, the negotiations start at a more realistic price and are often less time consuming. On the other hand, *Allied Security Trust* can save internal resources. *Allied Security Trust* works with a small team. Cooperating with a patent broker helps those team members focus on their core business. To evaluate the patents, *CreativE* works closely with a service provider specialized in patent valuation and patent evaluation. This service provider offers a legal check conducted by a patent attorney, as well as a market analysis conducted by a business specialist. By working with a service provider, *CreativE* is able to save personal resources and obtain an expert opinion for different areas of technology without building up internal competencies in this area.

Bundling patents to new patent portfolios is a continuous process and heavily dependent on negotiations and aggregation success, as well as the value-adding activities of the patent aggregating company. The actual allocation of the patent portfolio may differ from the initial planned allocation, because negotiations may not result in transactions.

3.2.3 Additional value adding activities

The degree of value added to a patent or a patent portfolio varies depending on the business model of the patent aggregating company. In general, the activity of patent aggregation itself is one of the most value-adding activities. Bundling patents leads to significant added value, because a patent portfolio has presumably a higher value than the sum of the single patents the portfolio is composed of (Chesbrough, 2006; Parchomovsky & Wagner, 2005). Patents are negative rights, and exclude third parties from using the invention, but they do not allow the patent owner to produce anything. Therefore, a single patent often has only a minor blockade function and can easily be circumvented.[4] Aggregating several single patents from one technology to a patent portfolio increases the overall value of each patent. Aggregating activities that enhance embryonic technologies create additional technical value. Already commercialized technologies are more predictable to evaluate and do not need further activities to market the products, whereas embryonic technologies require significant financial and managerial effort to create marketable products. Additionally, embryonic technologies that are successfully brought to the market foster the overall technological development.

Beyond the business model of patent aggregation itself and its associated value-adding activities, the most important activity that some patent aggregating companies perform is further development of the technology. *IgniteIP* for instance, focuses mainly on the commercialization of technologies and handles patents mainly as the right to the technologies. To commercialize the patents in the next process step, *IgniteIP* invests in further development. Mandated research institutions, universities, or companies conduct research and development. Results of the development are prototypes, components, or marketable products. Therefore, it transfers the embryonic technologies to technologies close to market entry. Accompanied by the technology development, *IgniteIP* also adds value to the patent itself by expanding the geographical scope of the patent, increasing the patent family, or drafting or adjusting patent applications.

Another important value-adding activity is the expansion of the existing patent portfolio. Due to the dynamics in the product market, as well as the patent market, new

---

[4] Some exceptions apply in the pharmaceutical industry. In this industry, a patent often covers one product, and the success of the product and the generated revenues heavily depend on the patent.

technologies or newly available patents supplement existing patent portfolios. *Via Licensing*, for instance, calls regularly for patents to update their portfolio and to secure clients access to the standardized technology. *Allied Security Trust* focuses on reducing the exposure of patent litigations for clients. Therefore, *Allied Security Trust* monitors the market closely and supplements its existing patent portfolio with patents that could be a threat to its members.

In the value-adding phase, patent aggregating companies prepare the communication documents for their patent portfolios. These communication documents can take several forms. *CreativE*, for example, aggregates patents that other companies use without having been granted a license. In this phase of the process, *CreativE* prepares all documents that show that products of certain companies use the patents without a license and of how they use the patents. The prepared documents serve as basis for the next phase and the negations and technical meetings with the infringing company. *Alpha Patentfonds*, for instance, prepares the sales documents in this phase. The sales documents contain all relevant legal, technical, and commercial information about the technology or the patents. *Alpha Patentfonds* uses the sales documents to approach potential customers, and they are of particular importance if the technology is offered to companies in other markets or other areas of application. Therefore, the documents have to be comprehensible for all types of buyers.

A different form of value adds for instance, *Pete Invest MedTech*. *Pete Invest MedTech* does not add value to the patents or the technology but creates value outside of the traditional licensing business. The licensing deal is already signed when *Pete Invest MedTech* steps in. The major value created by this patent aggregating company is indirectly related to the patents and is focused more on the global financing of the patent owner. In the value-adding phase, *Pete Invest MedTech* raises money from investors based on the patent and transfers the capital to the patent owners.

3.2.4 Exploitation of patents

In the last phase of the patent aggregating process, the patents amassed by the patent aggregating company are exploited. Patent aggregating companies, in general, have four different ways to utilize their patents:

- *Licensing* that can be subdivided into stick licensing and carrot licensing

- *Assigning* patents that can be subdivided into selling and realizing. Releasing means offering third parties the use of the patents without receiving financial compensation in return
- *Refinancing them at the capital market* or more specifically securitizing the cash flows resulting from a patent at the capital market.

Whereas the first two types of patent exploitation satisfy the demand for patents or the underlying technology, the last type satisfies the demand of financial investors and provides capital for the patent owner. Figure 11 illustrates the ways in which patent aggregating companies exploit their patents.

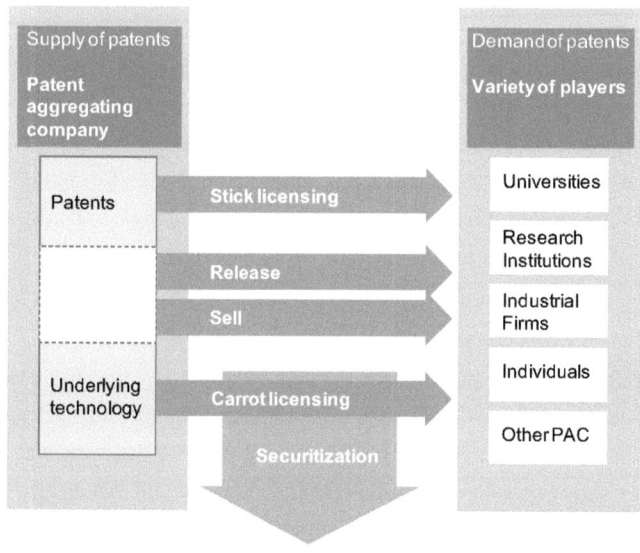

Figure 11: Patent exploitation options of patent aggregating companies

Out-licensing patents is an active patent management strategy of producing companies. Thereby, producing companies follow three main approaches in out-licensing: carrot (also known as opportunity) licensing, stick (also known as enforcement or assertion) licensing, and cross-licensing (Reinhardt, 2008). According to Lichtenthaler (2010), carrot licensing is a proactive approach and describes a technology transfer to licensees that have not used the technology yet. Stick licensing is a reactive approach and is based on actual infringements of the company's patents. The patent owning company detects these infringements and offers the infringer a license.

A long time part of corporate patent management, patent aggregating companies have adopted these commercialization strategies and applied the approaches of stick and carrot licensing to exploit their patents. For instance, *Patent Select* proactively offers interested parties licenses to its patent portfolio. The patents cover new technologies close to a commercialization stage. Additionally, knowledge is transferred to support and accelerate the final commercialization of the technology. Therefore, *Patent Select* satisfies the demand of new technology and the products of medium sized and large producing companies. In contrast, *CreativE* follows a stick licensing approach. Based on its strategy to acquire patents covering densely patented technologies only, *CreativE* detects companies that use the patents and proofs the infringement. Potential infringers are contacted to close licensing agreements. If the infringing company does not react after requests for technical meetings, *CreativE* takes the case to court. The results from negotiations are licensing agreements that allow the licensees to use the patents and give freedom to operate. Licensees are all kinds of producing companies applying the technology. *CreativE* does not transfer any additional knowledge.

A different approach of patent aggregating companies is to sell the patents to other companies. The offered legal rights can be single patents or patent applications, patent families, or patent portfolios and, depending on the objective of the buyer, transferred with or without additional knowledge. *Alpha Patentfonds*, for example, identifies potential buyers based on an intensive market research or its personal network and offers the legal rights for sale. Interested companies are mainly producing companies and other patent aggregating companies. They buy the patent portfolios and generate freedom to operate or access new technologies and enter new markets.

Some patent aggregating companies exploit their patent portfolios without gaining monetary returns. By giving non-monetary licenses to interested patent users, the covered technology is diffused and innovation in these areas is fostered. The users do not enter a financial commitment but rather enter a non-financial commitment. *Eco-Patent Commons*, for example, makes patents on sustainable technologies that bring environmental benefits available to anyone free of charge. The users are corporations, but single inventors and research institutions that work on innovation to protect the environment also use them.

The securitization of patents is another approach to exploit patents. In this approach the patent aggregating company does not directly exploit the patent, but it uses the existing licensing agreements and royalty streams for exploitation. Therefore, the patent aggregating companies do not act as middlemen, but rather meet the demands of capital market investors by creating financial instruments based on patents and royalty payments. *AlseT IP*, for instance, uses securitization of patents as a new asset class that enables financing of commercial products and technologies based on their future royalty income. *AlseT IP* issues bonds backed by the cash flow generated from the patents. Investing in these bonds, financial investors participate in the evolution of technology. The original patent owners, mainly SME or research institutions, receive an immediate cash flow. The licensee is unaffected by this transaction.

3.3 Strategies of patent aggregating companies

All analyzed patent aggregating companies amass patents, but they differ substantially regarding their reasons why they amass patents and in which business strategies they follow. Comparing the empirical data of the 27 case firms, 8 different strategic objectives why patent aggregating companies amass patents can be derived.

The eight objectives can be divided into two groups. The first group contains objectives to generate revenues by exploiting the patents. The second group contains objectives to serve members or customers of the patent aggregating company, or the society on a for- or non-profit basis. In this context, patents are only a means for achieving the objective. The exploitation of them follows more diverse objectives than sole revenue generation.

3.3.1 Basic strategy I: Generate revenues

The first group of patent aggregating companies that amass patents to generate revenues applies four different strategies of patent exploitation. One exploitation strategy follows a generic exploitation approach, and patent aggregating companies pursuing this strategy generate revenues from selling and out-license patents in various forms. For instance, *Intellectual Ventures* acquires large amounts of patents to exploit them in every possible way. The company sells patents, establishes licensing programs, enforces patents when they are infringed, or invests in additional R&D when the technology is still embryonic. The basis for *Intellectual Ventures's* business model is their ability to benefit from arbitrage. On the one hand, *Intellectual Ventures* has broad experience in patent transactions and a large network to detect good patents; on the other hand, the company is able to buy them at a lower price than it would cost to reinvent the acquired patents. Additionally, *Intellectual Ventures* benefits from the fact that patent portfolios have a greater value than a single patent.

Patent aggregating companies that stick license infringed patents follow another way to generate revenues from patents. For instance, *IP Navigation* pursues this very specific exploitation strategy and acquires only patents that are already infringed. The infringing company is contacted and negotiation is offered or infringement lawsuits are filed immediately. Therefore, the infringed patents are exploited through stick licensing.

Patent aggregating companies that carrot license technologies that are advanced during their ownership pursue another specific strategy to generate revenues from aggregated patents. *IgniteIP* pursue this strategy and focuses on promising technologies by acquiring the patents that cover these technologies. After investing in further R&D, *IgniteIP* follows a carrot licensing approach and out-licenses the advanced technology.

Patent aggregating companies also pursue a strategy that resembles the strategy of patent brokers, and first acquire patents or the exclusive right to exploit the patents and after that, sell them. *Patent Invest Fond*, for instance, pursues this strategy. *Patent Invest Fond's* strategy is to generate revenues from aggregating patents, bundle them to new portfolios, and sell these bundles at a higher price. Thereby *Patent Invest Fond* uses its extensive network and sells patents for other applications and industries.

3.3.2 Basic strategy II: Serve an objective

The second group of patent aggregating companies amasses patents as means to pursue four different objectives. One objective of patent aggregating companies is to offer attached companies an insurance against infringement. For instance, *RPX* monitors the market and detects patents that could be a litigation threat for attached companies. *RPX* aggregates the harmful patents before other producing or patent aggregating companies can acquire these patents and therefore, provides an insurance against patent litigation lawsuits for its attached producing companies.

Another goal patent aggregating companies pursue is to provide access to technology to a broad range of users to foster innovation and society. *Golden Rice PDP*, for instance, administers patents and technology that cover technology from the 'Golden Rice Project'. In this project, a strain of rice that contains pro-vitamin A was genetically engineered. By collecting suitable patents from several patent owners, *Golden Rice PDP* is able to neutralize licensing issues and make patents available free of charge for defined humanitarian research and use in developing countries by resource-poor farmers.

Certain patent aggregating companies amass patents to solve problems arising from patent thickets and to establish technology standards. For instance, *Via Licensing* amasses patents to provide producing companies access to the Advanced Audio Coding (AAC) technology. AAC is a compression and encoding scheme for digital audio. Compared to MP3, AAC achieves better sound quality and is the standard audio format for *Apple* products, such as iPhone, iPod, and iPad, as well as for Nintendo DSi, and PlayStation 3. *Via Licensing* has aggregated patents from *AT&T, Dolby Laboratories, France Telecom, Philips Electronics, LG Electronics, Microsoft, NEC Corporation, Nokia, NTT, Panasonic, Sony,* and *Ericsson* to provide access to essential patents for practicing the AAC technology.

Some of the patent aggregating companies provide capital to companies and use patents only as security. *Royalty Pharma*, for instance, is a patent aggregating company that uses patents for this purpose. An example is the agreement between *Royalty Pharma* and *Yale University* in 2000. In this deal, *Royalty Pharma* acquired the royalty streams of *Zerits®*, a drug for the treatment of HIV infection developed by *Bristol-Myers Squibb,* and securitized the royalty streams at the capital market. *Yale University* discovered a novel technology for HIV treatment, named d4T, and licensed this technology to *Bristol Myers Squibb* for the development of *Zerits®*. The US Food and Drug Administration (FDA) approved *Zerits®* in 1994. *Royalty Pharma* issued USD 115 million in debt and equity securities to fund the acquisition payment of the patents. *Royalty Pharma* provided *Yale University* an alternative source of capital and intended to use the future royalties to pay back the issued securities. The acquired patents served as security if the collected funds could not be paid back to the investors otherwise.

3.3.3 Eight business models of patent aggregating companies

The reason why patent aggregating companies amass patents determines the business model of the patent aggregating companies. All patent aggregating companies acquire patents from the original patent owners and give the owners some reward, but the reason why they aggregate patents and what they do with the patents differs in eight distinctive ways. Analyzing the 27 case companies, the following eight business models can be derived (alphabetical order):[5]

(1) Business model *patent acquisition company*: Aggregates large numbers of patents and technologies from a large number of original patent owners to low prices; bundles new portfolios or conducts further R&D; sells and stick or carrot licenses the patents to generate revenues and profit from arbitrage.

(2) Business model *patent enforcement company*: Aggregates patents that are already used and stick licenses them to generate revenues.

(3) Business model *patent incubating fund*: Aggregates promising, often embryonic technologies, and the patents that cover the technologies; enhances technology by conducting further R&D; and sells or carrot licenses technology to generate revenues.

(4) Business model *patent trading fund*: Aggregates large numbers of patents from a large number of original patent owners to low prices; bundles new portfolios; and sells new bundles within but also across industries to generate revenues.

(5) Business model *defensive patent aggregator*: Aggregates patents used by its attached companies that could create a litigation threat for the attached companies.

(6) Business model *non-commercial patent aggregator*: Aggregates patents and technologies from several patent owners often without giving direct monetary rewards; offers the patents to a broad range of users without charge.

---

[5] The terms used to name the eight business models build on prior research and internet documents. In prior research and in internet documents, several denominations and terms are used overlapping, unlimited, and/or synonymous. For instance, von Scheffer (2008), p. 5 mentions the terms 'patent incubating funds' or 'patent trading funds'. Millien and Laurie (2008), p. 54 use the terms 'patent licensing and enforcement funds' as well as 'IP acquisition funds'. Yanagisawa and Guellec (2009), p. 11 use the term 'defensive patent aggregation funds'. Except for the concept of patent pools, which is described for example, in Aoki and Schiff (2008), a distinction and definition of the terms is lacking so far.

(7) Business model *patent pooling company*: Aggregates large amounts of patents covering certain technologies from several original patent owners; offers a single license to all patents of different owners.

(8) Business model: *royalty monetization company*: Aggregates patents that are already licensed out and produces steady royalty streams as security for capital provided to original patent owners.

Figure 12 summarizes the business models of patent aggregating companies, assigns them to one of the two general strategies, and displays which company shows which business model.

Patent aggregating companies acquire large amounts of patents to….	
…generate revenues (basic strategy I)	…serve an objective (basic strategy II)
• **Patent acquisition company: Selling, stick- and carrot licensing – applying arbitrage strategies** (Coller Capital, Intellectual Ventures, Techquity) • **Patent enforcement company: Stick-licensing infringed patents** (Acacia, Alliacense, CreativE, Fergason Patent, IP Navigation, Papst, Rembrandt) • **Patent incubating fund: Carrot licensing refined technologies** (IgniteIP, IP Holdings, Patent Select) • **Patent trading fund: Mainly selling patents** (Alpha Patentfonds, Patent Invest)	• **Defensive patent aggregator: Offering insurance against infringement lawsuits to members** (Allied Security Trust, OIN, RPX) • **Non-commercial patent aggregator: Offering technology access to a broad range of users to foster innovation and society** (Golden Rice, Eco-Patent Commons) • **Patent pooling company: Offering access to standards** (MPEG, Sipro Lab, Via Licensing) • **Royalty moentization company: Offering alternative source of capital** (AlseT, Capital Royalty, Pete Invest MedTech, Royalty Pharma)

Figure 12: Business models of patent aggregating companies and their strategies

3.4 Summary

During the last two decades, companies that do not produce physical goods have emerged as buyers in the market for patents and technologies. As little is known about these patent aggregating companies, an explorative analysis is conducted on companies that are visible and have a certain, observable, track record in patent aggregation. This analysis confirms that patent aggregating companies are a young phenomenon, since the first company started aggregating activities in 1996. Patent aggregating companies are established based either on pre-existing activities or by entrepreneurs, patent professionals, or financial institutions.

Patent aggregating companies are very heterogeneous regarding their number of employees, their patent portfolios, and their capital endowments. Whereas most companies operate with a number of employees around 57, one company exceeds all companies regarding size and asset under management. Patent aggregating companies follow a general process to amass patents. This process consists of the four phases: selection, structuring, value adding, and exploitation. Depending on the patent aggregating company, each phase is of different importance. They amass either the sole legal right of exclusion or patents, in addition to technology and knowledge.

Two basic strategies regarding why patent aggregating companies amass patents can be identified: to generate revenues and to serve an objective. Four different ways of how they generate revenues can be recognized: (1) through a broad exploitation strategy and selling, stick and carrot licensing; (2) through stick licensing of infringed patents; (3) through carrot licensing of refined technologies; (4) through selling patents. Also four different objectives for which patent aggregating companies use patents as means to pursue these objectives can be identified: (1) they offer an insurance against infringement lawsuits to members; (2) they offer technology access to a broad range of users to foster innovation and society; (3) they offer access to standards; (4) they offer an alternative source of capital. The identification of the eight different strategies of patent aggregating companies answers research question 1: *Why do patent aggregating companies build up large patent portfolios?*

From the eight different strategies, eight distinctive business models are identified: patent acquisition companies; patent enforcement companies; patent incubating funds; patent trading funds; defensive patent aggregators; non-commercial patent aggregators; patent pooling companies; and royalty monetization companies.

4 Potentials offered by patent aggregating companies

As an intermediary step in examining whether patent aggregating companies are an option for producing companies, the potentials offered by patent aggregating companies are analyzed. The focus is on potentials that patent aggregating companies can provide for producing companies that want to leverage their patent portfolio, and not on how the patent aggregating companies can be beneficial for patent buyers or licensees, capital market investors, or the economy or society. Potentials can be found either internally, that means in the producing company itself, or externally in the market environment (Figure 13).

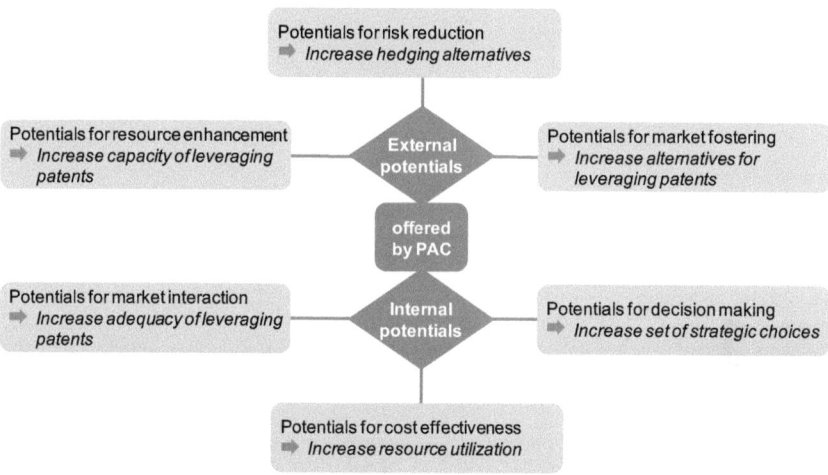

Figure 13: Overview of external and internal potentials

External potentials describe opportunities created by patent aggregating companies for the producing company in the environment. The producing company can benefit from them by utilizing patent aggregating companies but cannot actively influence them. This is in contrast to internal potentials that lie in the producing company itself.

In the following, case studies and examples from literature demonstrate how patent aggregating companies can realize the external and internal potentials.

4.1 External potentials offered by patent aggregating companies

External potentials offered by patent aggregating companies are potentials for risk reduction, market fostering, and resource enhancement. Patent aggregating companies offer producing companies new opportunities to hedge risks. Not only do hedging opportunities have to exist but leveraging options and an environment also have to be present. In addition, the producing company has to seize the opportunities.

4.1.1 Potentials for risks reduction

External potentials for risks reduction can be realized through hedging of R&D risks and hedging of enforcement risks (summarized in Figure 14). Realizing risks reduction increases the hedging alternatives for producing companies.

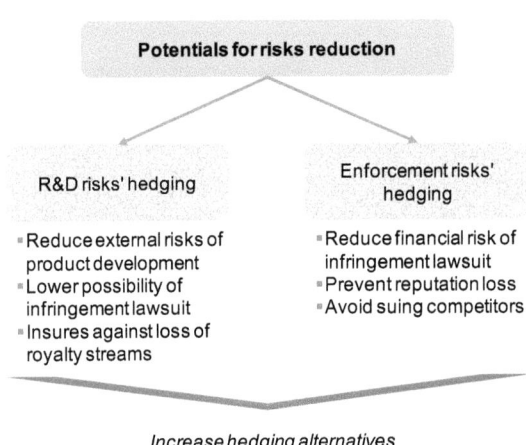

Figure 14: External potentials for risks reduction

R&D projects are considered typically to be high-risk projects (see Gassmann, Kobe, & Voit, 2001), and the commercialization of R&D is exposed to two main types of risk: failures related with R&D and the results of R&D, and failures related with the enforcement of the legal right on the invention. Stevens and Burley (1997) found that across most industries, 3,000 raw ideas are required to produce one substantially new commercially successful industrial product. Developing in the wrong direction, therefore, is costly. Another risk connected with R&D is the inadvertent use of other companies' patents. Even if the invention and the patent are commercialized, for instance, via out-licensing, R&D results could still remain subject to risk because licensees could fail to pay the royalties. If the patents are enforced, the producing company faces risks related with infringement lawsuits. Patent aggregating companies can reduce risks from R&D, as well as the risks from enforcing the patents.

R&D risks hedging

Patent aggregating companies can reduce producing companies' external R&D risks by taking over the tasks of further development and commercialization and providing rents of innovation immediately to the producing company. Figure 15 gives an overview of what can go wrong in the process of product development. External risks of R&D can be divided into market and technology risks. These two risks describe the financial uncertainty of an innovation regarding the achievement of an attractive financial return (Rogers, 2003).

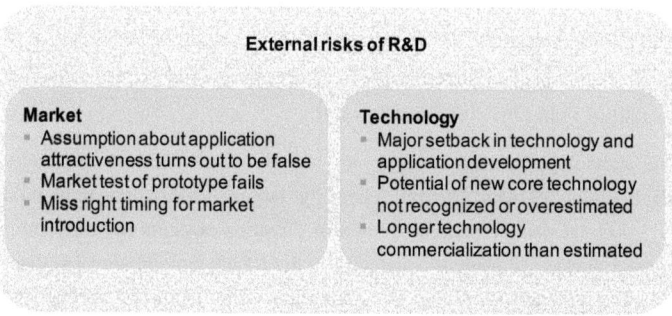

Source: Adapted from Becker (2003).

Figure 15: External risks of R&D

By involving patent aggregating companies and selling the patents and the covered technology, producing companies can achieve an attractive financial return before the product is actually commercialized. Therefore, the problems listed in Figure 15 are no longer prevalent and transferred to the patent aggregating company.

Techquity acquires patents and the covered technologies from corporate sellers as SMEs and MNEs. Thereby, *Techquity* focuses on technologies that have yet to be commercialized in products, but have a proven technical value in applications and benefit future markets. The original patent owners receive a rent from their innovation before it is commercialized in products, without carrying the market and technology risks of further product development. Furthermore, they receive a back license. However, both facts, the risks transfer, as well as the back license, may lead to lower rents of innovation but protect the original patent owner from a total loss of financial returns in case of any undesired development.

For producing companies, patent aggregating companies can lower the risk of being involved in infringement lawsuits as a defendant. This is especially true in industries characterized by large numbers of patent applications, overlapping sets of patents, and patent thickets, as producing companies constantly face the threat of, often inadvertently, unauthorized use of patents. Even with conducting serious patent monitoring and the freedom to operate analyses, producing companies are not always able to detect all relevant patents and may develop new products that inadvertently infringe patents of competitors or other companies. In some technological areas, where many companies face the same problem, the involvement of a patent aggregating company can solve this problem. Assigning own patents to a patent aggregating company allows producing companies access to patents of other companies that cover the same technological area. This organized cross-licensing procedure reduces the risk of infringing patents and of being involved in a lawsuit.

> Even a young technology, patent thickets are already observable in the Radio Frequency Identification domain (RFID). The fifty largest RFID innovators hold approximately 3,000 patents. To prevent a flood of litigation lawsuits resulting from this growing patent thicket, 20 leading companies in the RFID domain, amongst them *3M, France Telecom, Hewlett-Packard, LG Electronics,* and *Motorola,* formed the RFID Consortium and hired *Via Licensing* to establish a patent pool and submit a business plan to the US Department of Justice, which reviewed the patent pool and the RFID Consortium in order to ensure its arrangement does not threaten any antitrust laws.[6]

Patent aggregating companies can insure producing companies against loss of royalty streams through prepaying future royalties and taking over the risk of total (or partial) failure of the licensee. Fishman (2003) identifies two scenarios in which this insurance

---

[6] *Via Licensing* has established the patent pool. In 2009, patent pool administrator *Sisvel Group* took over the administration and licensing of the UHF RFID patent licensing program.

scheme could be applied. The most straightforward case is that a licensee defaults to pay the royalty. Reasons for this could be insolvency of the licensee, failure of the product in the market, or obsolete technology. Another insurance scenario is an invalidated patent. If the patent is invalidated, the licensee is discharged from paying financial commitments. Patent owners can avoid suffering losses caused from unpaid royalties by reassigning patents to patent aggregating companies and receiving discounted and adjusted royalty payments.

Enforcement risks hedging

Patent aggregating companies can reduce the financial risks of infringement lawsuits, in which a producing company acts as plaintiff. If a patent owned by a producing company is infringed, the patent owner has few alternatives other than patent litigation. In general, patent litigation is usually lengthy, it is always very expensive, and it is often unsuccessful.

On average, patent litigation lawsuits in the US last about two years (Barry et al., 2010) but can be many times longer.[7] For instance in 2007, an infringement lawsuit of *Microsoft* vs. *Eolas Technologies* and *University of California* was settled after eight years (Bloomberg News, 2007).[8] Patent litigation lawsuits in the US cost on average USD 3 million to 10 million (Towns, 2010).[9] Additionally, results of patent litigation at the appellate level show that patentees only won some 25% of the cases (Janicke & Ren, 2006). These numbers show that enforcing patents is a risky business.[10]

---

[7] The average duration of patent litigation lawsuits in Europe depends on the country. Whereas on average, litigation lawsuits in Germany are shorter than in the US (1–1.5 years), litigation lawsuits in France are only slightly shorter (1.5–2 years). In Italy (3 years) and England (2–3 years to finish hearings and to come to a trial), the average duration of litigation lawsuits is even longer than in the US (Aoki and Hu, 2003).

[8] "*Microsoft* said Thursday that it had settled an eight-year patent dispute that resulted in a USD 521 million jury verdict against it. Terms of the accord were not disclosed. The dispute centered on a feature within *Microsoft's* Internet Explorer Web browser that allows embedded links. The patent is owned by the *University of California* and licensed to *Eolas Technologies*, a closely held company formed by a university researcher, Michael Doyle. " (Bloomberg News, 2007)

[9] The costs for patent litigation lawsuits in the US vary depending on the amount of money at risk. In 2011, for patent infringement suits with less than USD 1 million at risk, the median costs are USD 600,000. In patent infringement suits dealing with USD 1 to 25 million at risk, the median costs are USD 2 million and increase further to more than USD 5 million if the value in litigation exceeds USD 25 million (AIPLA, 2011).

[10] This is aggravated by the fact that in the US, non-specialist judges or juries consisting of lay people often decide the outcome of infringement lawsuits. The assessment of infringement requires technological knowledge and is complex process. Therefore, the outcome of a litigation lawsuit is difficult to predict (Luman III and Dodson, 2006).

Assigning (potentially) infringed patents to patent aggregating companies can reduce the financial risks resulting from the lawsuit.

> *Papst Licensing* takes over the enforcement risks of mainly European companies operating in the US market. After a thorough analysis of the potentially infringed patent, *Papst Licensing* buys the patent from the original patent owner and usually compensates the original patent owner through a sales price that consists of an upfront payment and a variable success related component (back end). Therefore, the original patent owner already receives a payment independent of an enforcement result and participates in case of a monetization success.

Additionally, patent aggregating companies can reduce risks that result only indirectly in financial losses. Producing companies have two major reasons for not being involved in patent enforcement cases. First, in some communities, the enforcement of patents and the patent system in general are considered to hinder innovation (e.g., The Economist online, 2010). Therefore, companies perceived as very innovative can reduce the risk of reputation loss resulting from being involved in patent litigation and still receive a rent for their innovation through selling infringed patents to a patent aggregating company. In this case, patents are enforced but the original patent owner is not involved. Second, producing companies operating in markets with oligopolistic structures often abandon the option of patent enforcement. Due to the small number of competitors, a litigation case could stir a flood of lawsuits and end in a zero sum game. Assigning infringed patents to a patent aggregating company leaves the original patent owner out of the lawsuit while receiving a rent for the innovation.

4.1.2 Potentials for market fostering

External potentials for market fostering comprise liquidity, market clearing, and innovative business models, as summarized in Figure 16. Offering these potentials, patent aggregating companies can increase the alternatives for leveraging patents for producing companies.

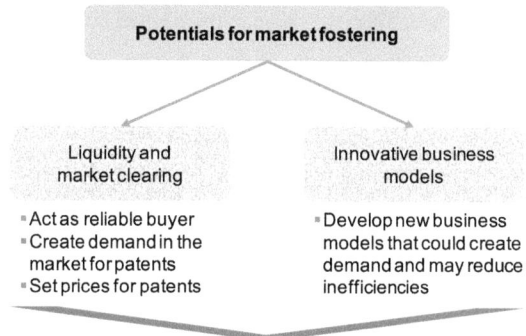

Figure 16: External potentials for market fostering

The options for leveraging producing firms' patent portfolios have increased since patents are viewed as valuable and separate from a company's core business. The external exploitation of patents is executed in market for patents and technologies (Granstrand, 2000) and patents (Gambardella et al., 2007), but still this market lacks transparency, liquidity, and information symmetries. As active players, patent aggregating companies can provide liquidity and support the pricing of patents. As the current patent market environment can be classified as a period of trial and error (Malackowski, Cardoza, Gray, & Conroy, 2007), new business models emerge and vanish. Patent aggregating companies already have experience in the market for patents and technologies and can enhance the success and the reliability of new business models.

Liquidity and market clearing

As buyer in the market for patents and technologies, patent aggregating companies can enhance the liquidity of patents and foster the development of the market for patents. Liquidity of an asset is defined as "the time and costs associated with the transformation of a given asset position into cash and vice versa" (Jorion, 2009, p. 607). In other words, liquidity refers to the ability to unwind a position on short notice without influencing the market price. Therefore, liquidity should encompass the following three components: (1) time required to sell an asset; (2) transaction costs incurred when selling the asset; (3) the degree of uncertainty in the liquidation value of the asset (Murphy, 2008). Due to the high specificity of patents, the limited context a patent can be applied, and potential circumvention opportunities, it takes a long time span and high search and transaction costs to transform patents into cash. Therefore, patents are illiquid assets. Patent aggregating companies, appearing as regular buyers of vast amounts of patents, can lead the patent market from a search market with multiple decentralized buyers and sellers that are not aware of each other, to a centralized market (McDonough III, 2006; Shrestha, 2010). Hence, patents become a more of a liquid asset through a centralized demand and the coordinating function of patent aggregating companies.

Additionally, patent aggregating companies can clear the market by equalizing prices (McDonough III, 2006). In markets with information asymmetries, participants cannot assess the value of a patent and therefore, are not able to set a consistent price. That leads to market failure. Patent aggregating companies act as buyers and sellers in the market and have more experience in pricing and better access to information. Therefore, patent aggregating companies are more able to set market clearing prices.[11]

> One example of a company with vast experience often named as the smartest buyer in the market is *Allied Security Trust* (Hetzel, 2010). To buy a patent, *Allied Security Trust* needs the commitment of its members. The members also independently value the patents and determine the amount they contribute to the bid. Based on the accumulated patent management and patent valuation experience of high-technology companies, *Allied Security Trust* determines the price.

Innovative business models

By developing new business models, either separate to the aggregating activities or additional to them, patent aggregating companies can create additional demand and provide additional expertise and resources. Therefore, they may help to reduce market inefficiencies. According to the EPO, OECD, and UKIPO (2006), "the IP marketplace is nowadays in a probe and learn period where the number of intermediaries is rising" (p. 1). As many firms are not able to overcome market imperfections on their own, new business models that bring together supply and demand emerge and vanish. Therefore, patent aggregating companies, already working in the market for patents

---

[11] McDonough III (2006) alludes to the opportunity that large buyers of patents can use market imperfections to benefit themselves by setting prices too low. He indicates that this problem is likely to abate over time as buyers and sellers become more experienced in setting transaction prices.

and identifying new business opportunities, can develop new models and "make one step forward towards the development of a market for IP transfers ...[and]... contribute to the maturation of the IP market" (EPO et al., 2006, p. 1).

4.1.3 Potentials for resource enhancement

By utilizing patent aggregating companies, producing companies can realize external potentials for resource enhancement through access to human and financial resources, as well as through networks. Figure 17 summarizes the potentials for resource enhancement. Having access to external resources can increase the capacity for leveraging patents.

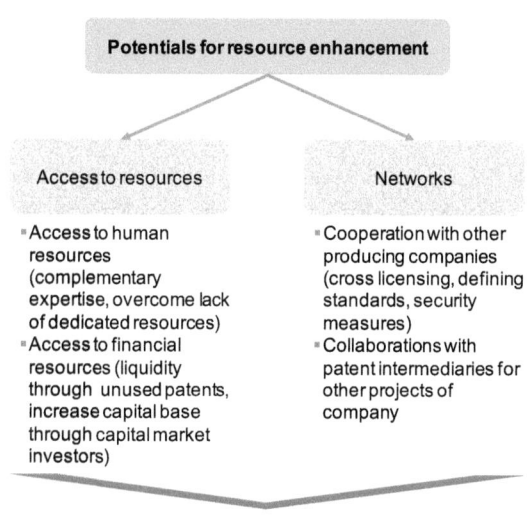

Figure 17: External potentials for resource enhancement

Companies' own resources can enable them to achieve competitive advantage and lead to superior long-term performance. Literature distinguishes between physical, organizational, financial, and human resources (Barney, 1991) and knowledge (Kogut & Zander, 1992). Leveraging patent portfolios successfully requires resources and competencies of producing companies. Firms can release resources internally for patent leveraging projects and develop their own competencies, or seek assistance from third parties (Nambisan & Sawhney, 2007; Sapsed et al., 2007), for instance, patent aggregating companies.

Access to resources

Patent aggregating companies can enhance access to external resources. According to Arora et al. (2001b), the external acquisition of resources is of growing importance in companies' strategic options. Patent aggregating companies enable producing companies to access to financial and human resources.

Human resource. On the one hand, patent aggregating companies complement companies' internal resources with complementary expertise (Morgan & Crawford, 1996; Nambisan & Sawhney, 2007), and producing companies have to build up fewer internal resources to leverage their patent portfolio successfully (Shohert & Prevezer, 1996). Complementary expertise provided by patent aggregating companies is mainly highly specific resources that are also highly cost intensive, for example, patent lawyers.

> *Alpha Patentfonds* offers producing companies industry and market expertise that increases opportunities to leverage the patent portfolio. Companies that sell patents to *Alpha Patentfonds* often have their own patent departments and therefore, are able to evaluate the quality of the patent, as well as the technological applicability. Despite the patent management expertise, the original patent owners do not have any expertise in external patent transactions, limited resources to sell abandoned patents, and only limited access to potential buyers. *Alpha Patentfonds* has the necessary access to buyers and the negotiation and implementation expertise that leads to a successful completion of external patent portfolio leveraging activities.

Additionally, many producing companies follow an ad hoc approach for exploiting patents externally and therefore, lack dedicated resources (Lichtenthaler & Ernst, 2008b). Patent aggregating companies can help producing companies to overcome this human resource bottleneck by providing additional human resources.

> *CreativE* acquires most patents from companies that do not have their own patent department or own patent lawyers, such as SMEs, research institutions, or single inventors. Therefore, these companies are not able to detect infringements and companies that potentially use their patents to enforce the patents. By acquiring potentially infringed patents, *CreativE* provides indirect human resources to these companies through monitoring the market for potential infringers, analyzing products

of competitors for using the patents, and finally employing specialized and experienced patent lawyers to enforce the patents.

Financial resources. SMEs and privately held producing companies have constraints regarding the financing of R&D because they cannot access capital through normal capital market instruments. Additionally, the access to debt capital borrowed by banks has become more difficult through increasingly strict regulations like Basel II (Bessler, Bittelmeyer, & Lipfert, 2003). Patent aggregating companies can help producing companies to increase liquidity and access capital markets without the companies becoming involved with traditional debt and equity instruments.

Patent Select aggregates patents from SMEs, single inventors, research institutions like universities, but also from MNEs. Often the original patent owner does not have the financial resources to commercialize the results from own R&D or holds patents that diverge from the general strategic direction of the company. *Patent Select* acquires patents covering promising technologies not used by the original patent owner and commercializes them. Therefore, the original patent owner receives liquid assets that can be reinvested in company activities, for example, R&D. The advantage of *Patent Select's* business model is that it taps capital market investors. An investment fund structure finances the funds used to acquire the patents, as well as the enhancement. Investors can profit from patents as new asset class and risk diversification opportunities.

Networks

Patent aggregating companies can enhance the cooperation between producing companies and establish interfirm networks. Cooperation and interfirm networks can help to increase the supply side innovations, patents, and products but can also act as a common defensive shield. The interfirm networks are based on formal long-term agreements. They can be a source of strategic advantage because they can facilitate entry to a new market, share costs and risks, or help to establish a new technology standard in a particular market or industry. Patent aggregating companies can optimize the different collaborations and interfirm networks. For example, in high-technology industries, such as consumer electronics, telecommunications, or information technologies, a proliferation of patents is observable and may result in patent thicket (Aoki & Schiff, 2008). To secure a wide adoption of innovative technologies for products, which also represents the commercial interests of patent owners, companies have to collaborate to give access to overlapping and necessary patents.

MPEG LA administers patents of a video compression technology. The technology reduces the number of bits in a file. Based on the lower number of bits, videos can be transmitted faster and made available over lower bandwidth carriers. *MPEG LA* offers a 'one-stop shopping' for the licenses necessary to produce MPEG-2 products. The license offers non-discriminatory access to all essential patent of the MPEG standard. To offer these licenses producing, companies, such as *Alcatel Lucent, Canon, Columbia University, France Télécom, Fujitsu, Hitachi, Mitsubishi, Philips, Bosch, Samsung,* or *Toshiba* collaborate and pool their patents that *MPEG LA* administers. The network and collaboration agreement is expanded regularly to update the patent pool, and new licensors and essential patents are included. The patent aggregating company *MPEG LA* has established the interfirm network and ongoing collaboration of the original patent owners.

Additionally to enhancing interfirm networks, patent aggregating companies can promote access to indirect contacts with patent intermediaries or patent service providers and therefore, foster future patent leveraging activities. Patent aggregating companies often employ third parties for their own patent aggregation and patent commercialization activities. Based on this network of experts, producing companies have references for certain tasks and can benefit from the selection and employment done by the patent aggregating companies without having the search and selection costs and the risk to employ unskilled third parties.

4.2 Internal potentials offered by patent aggregating companies

Besides realizing the external potentials of the market and industry environment in a macro-context, patent aggregating companies can also offer internal potentials by helping to perceive leveraging opportunities, organizing micro-processes, and expanding the scope of action. Internal potentials offered by patent aggregating companies are potentials for market interaction, cost effectiveness, and decision-making.

4.2.1 Potential for market interaction

Internal potentials for market interaction include market understanding and opportunity identification. Figure 18 provides a summary of the potentials for market interaction. Realizing these potentials raises the adequacy of the offered patents and the market orientation of the transactions.

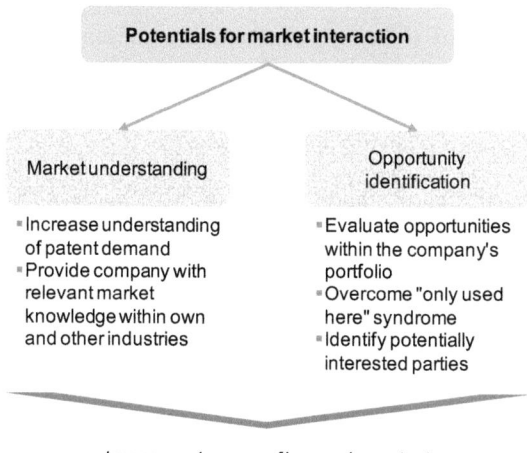

Figure 18: Internal potentials for market interaction

Leveraging patents through external patent exploitation is a much more complex task than commercializing goods on product markets (Callon & Muniesa, 2005). Prior studies have identified that major managerial difficulties are the identification of external patent exploitation opportunities and the determination of transaction prices (e.g., Arora et al., 2001a; Lichtenthaler, 2007a; Morgan & Crawford, 1996; Tschirky & Escher, 2000). Additionally, reliable data about the size, structure, and demand in the market for patents is not available (Yanagisawa & Guellec, 2009). Patent aggregating companies can serve as an interface to the market place because they have information and experience advantage over producing companies that seldom act in the market of patents or have separate units for patent management and for instance, technology scouting.

Market understanding

Patent aggregating companies operate regularly in the market for patents and therefore, can provide producing companies access to market data. With this data, producing companies are able to understand the market demand for patents regarding its structure, size and capacity, potential growth, and competition. Understanding the patent demand is a prerequisite for developing patent portfolio leveraging strategies. Patent aggregating companies can supply information about what potential patent buyers or licensees need and want and, if the patent owner participates in future royalties, to a certain degree, what they are willing to pay. Producing companies can use this information to determine a patent portfolio leveraging strategy and to adjust the innovation strategy to the market demand. Providing market information, patent aggregating companies are also able to apply their industry spanning network and knowledge and provide producing companies with market knowledge and application potentials of the patents in other industries.

Opportunity identification

Many producing companies still leverage their patent portfolios mainly internally. Often they own technical solutions for certain problems, innovative technologies, or patents but they face the difficulty of identifying possible applications in their industry and in completely different industries from the firm's own product business (Lichtenthaler, 2005). Additionally, managers fear that they might give away 'corporate crown jewels' when they sell or exclusively out-license patents (Kline, 2003). If patent aggregating companies are going to acquire certain parts or the entire patent portfolio, they can conduct an analysis of the producing firm's patent portfolio, as well as support producing companies by structuring and enriching internal analysis, therefore, reducing the difficulties of opportunities' detection.

> *IP Holdings* invests in, develops, incubates, and assists in the commercialization of novel and promising technology. The patent aggregating company particularly emphasizes the development and protection of patents. Additionally, *IP Holdings* offers management and audit services. In the patent aggregating process, *IP Holdings* audits patent portfolios of SMEs, universities, and other patent owners; identifies core, non-core and obsolete technologies; and based on this, detects external exploitation opportunities. The idea incubator acquires breakthrough technologies and disruptive innovation from life science or electrical engineering and these are further advanced and commercialized.

Patent aggregating companies can help to overcome the cultural problem, which is still immanent in many producing companies, of the 'only-used-here' (OUH)-syndrome (Boyens, 1998). The OUH-syndrome is defined as an attitude to the external exploitation of knowledge that is more negative than an ideal economic attitude would be (Boyens, 1998). A consequence of the OUH-syndrome is that companies cannot benefit fully from all patent portfolio leveraging options, for instance, companies fail to establish industry standards based on their own technologies or they are not able to gain access to patents in bilateral contracts as cross-licensing agreements (Lichtenthaler & Ernst, 2006). According to Lichtenthaler and Ernst (2006) the main reasons for the OUH-syndrome are the fear of strengthening competitors, a lack of experience with external patent exploitation, and the legal and organizational effort of external patent leveraging activities. Patent aggregating companies can provide track records of successful transactions that build up trust and reduce companies' internal inhibitions based on inexperience. Through shifting the complete transaction process to the patent aggregating company, the producing company can save organizational and legal effort. In the contracts, producing companies can exclude certain companies as receivers for their technology or their patents. These contractual agreements can reduce apprehensions regarding competitors' strengthening.

4.2.2 Potentials for cost effectiveness

Internal potentials for cost effectiveness comprise organizational learning and transaction costs. Figure 19 gives an overview of the potentials for cost effectiveness. Based on the competencies of patent aggregating companies, producing companies can increase their resource utilization with regard to patent portfolio leveraging activities.

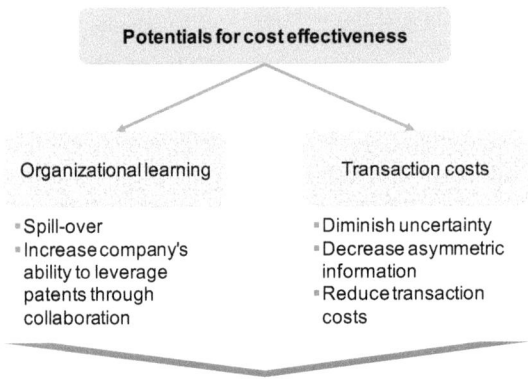

Figure 19: Internal potentials for cost effectiveness

For most producing companies, transactions of patents, and therefore the leveraging of the patent portfolio, are difficult to realize. According to Tietze (2011), the main obstacles of efficient transactions are uncertainty and information asymmetry, – actually leading to – high transaction costs. Tietze (2011), remarks that, "these three are intertwined and hardly can be distinguished clearly" (p. 60). As mentioned before, many producing companies follow an ad hoc approach for exploiting patents externally and lack dedicated resources (Lichtenthaler & Ernst, 2008b). Therefore, patent transactions and patent leveraging activities are characterized by high transaction costs.

Organizational learning

By developing internal competencies of patent transactions and patent portfolio leveraging activities, a firm may reduce its transaction costs (Cohen & Levinthal, 1990; Lane, Koka, & Pathak, 2006). Patent aggregating companies already have the necessary competencies in the market for patents and technologies and therefore, can help producing companies to build up internal competencies by learning from them (Kale, Dyer, & Singh, 2002; Lichtenthaler & Ernst, 2008b; Silverman, 1999).

> The main transaction partners of *Patent Select* are research institutions and SMEs. *Patent Select* acquires the patents to advance and subsequently out-licenses them. The original patent owner is not only compensated by lump sum payments or shared royalties, but also involved in the enhancement of the patents. For instance, *Patent*

> *Select* acquired a technology from a German SME that already had strong competencies to commercialize the technology in the domestic market but lacked market knowledge for specific international markets. Working closely with *Patent Select* and its network of international sales specialists, the SME was able to build up international marketing competencies and apply the new competencies for later products.

According to Lavie (2006) and Lichtenthaler and Ernst (2008b), the patent owner may generate inbound spillover rents from non-shared resources of the patent aggregating company. Therefore, transferring patents to a patent aggregating company and working with it may help the patent owner to realize learning effects that exceed the resources a patent aggregating company provides (Howells, 2006)

Transaction costs

Uncertainty. By using their particular technical, legal, and commercial expertise and their knowledge of supply and demand, patent aggregating companies help to reduce uncertainty. Uncertainty regarding the quality of patents (Gans et al., 2008), the transaction process (Lichtenthaler, 2004), the applicability in new environments (Caves et al., 1983), and especially the value of the patents and the technology (Gambardella et al., 2008; Scherer & Harhoff, 2000) hinders market transaction and prevents patent owners from successfully exploit patents externally. The value of a patent is not a fixed parameter, but it is among other things dependent on the actual utilization of the patent (Hall & Ziedonis, 2001). Therefore, the transaction price is difficult to determine, and often the asking price of buyer and seller differs substantially. Literature offers a broad range of valuation approaches,[12] but all approaches struggle with the same phenomenon that confronts the actors: uniqueness of a patent (Granstrand, 2000) and the difficulty comparing the traded patents (Parr & Smith, 2008), an elementary precondition for finding a transaction price.

Based on the experiences of past transactions and data collected in these transactions, patent aggregating companies are able to value patents and reduce the uncertainty regarding the transaction price.

> *IP Bewertungs AG*, the patent manager of the *Patent Select*, has developed a valuation method for patents on the basis of experience and on data of patents already priced and

---

[12] An intensive overview of the different patent valuation approaches can be found in Parr and Smith (2008).

traded gathered from different sources, such as expired license agreements, remunerations of employees' inventions, and patent sales (e.g., out of liquidations). With this data and academic literature,[13] value-indicators were located and significant correlations between indicators and values identified. The result is a valuation method that follows a market-approach with value indicators that was certified on customers' request by chartered accountant *KPMG* in February 2004. *Patent Select* applied the market-approach with value indicators to value the targeted patents and as basis for the transaction price negotiations.

Asymmetric information. Patent aggregating companies can reduce information asymmetries between the original patent owner who offers patents for sale or out-licensing and potential patent buyers respective licensees. Today's technologies and inventions are complex and often difficult to evaluate for companies and persons that have not participated in the development of this invention. Hence, the economic value of a patent is difficult to estimate for outsiders and asymmetric information between patents buyers and patent sellers exist. Based on the asymmetric information, a classical 'lemon market' (Akerlof, 1970) arises.[14]

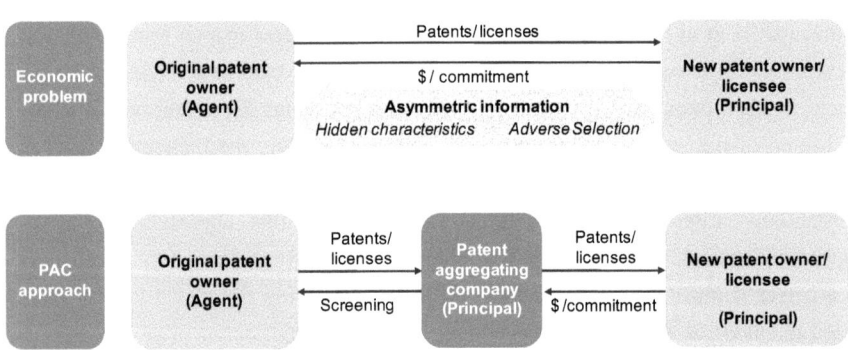

Figure 20: Principal-agent problem and patent aggregating company approach

---

[13] The selection of the patent value indicators is based on the following empirical studies: Narin, Noma, and Perry (1987); Lanjouw and Schankerman (2001), Lanjouw and Schankerman (2004); Harhoff, Scherer, and Vopel (2003a); Harhoff and Reitzig (2004); Reitzig (2004).
[14] Is the patent market a market for lemons, the seller would sell low value patents. The buyer knows that and only buys at low prices. Sellers of patents with high economic value could sell patents only to a low price and therefore, they would leave the market. In this case, the market for patents would be small and populated by low value patents.

Analyzing the market for patents in the context of principal-agent problems (Eisenhardt, 1989b; Jensen & Meckling, 1976), the original patent owner (agent) has superior information about the economic value of the patent (hidden characteristics). Resulting from this lack of information, the potential patent buyer (principal) might make an undesired decision and buy an unwanted patent (adverse selection). To reduce information asymmetries and increase the chances of patent transaction for the original patent owner, the patent aggregating company steps in as principal and acquires the patent. The patent aggregating company invests money in evaluating the patent, conducting a due diligence, and buying the patent, therefore, reducing the information asymmetries through the idea of screening (Figure 20). Through investing financial and human resources, the patent aggregating companies additionally signals the quality of the patents.

Patent Select reduces asymmetric information resulting from hidden characteristics and adverse selection through a resource intensive screening process (Figure 21).

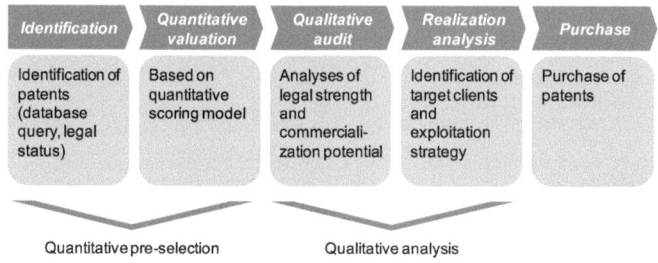

Source: According to Lipfert and von Scheffer (2006).

Figure 21: Screening and selection process of Patent Select

In general, *Patent Select* aggregates embryonic technologies from SMEs or research institutions. Due to the novelty of the technology, information asymmetries regarding the economic value of the patents are large and potential patent buyers cannot assess the quality and the potential. Therefore, the original patent owners are not able to sell the technology and the resulting patents on the market for patents and technologies. *Patent Select* reduces the problem of hidden characteristics through a five-step selection process: identification, quantitative valuation, qualitative audit, realization analysis, and purchase. Investing resources in the screening and taking risks in the acquisition, *Patent Select* signals quality to other potentially interested parties in the market for patents and technologies.

Transaction costs. Patent aggregating companies exploit the amassed patents by satisfying the demand for patents or the underlying technology (see section 3.2). Therefore, patent aggregating companies act as intermediaries, matching the supply of patents and the demand for patents or technologies from corporate buyers or other patent aggregating companies. Based on this intermediary function, patent aggregating companies can reduce transaction costs. The patent aggregating company in an intermediary form is located between the traditional choices of Institutional Economics hierarchy and market.[15] As an alternative to the direct transaction of patents, they serve as a governance mode to execute risky transactions. Patent aggregating companies are highly specialized and have comprehensive market knowledge. Therefore, they can also reduce operative costs, through for instance, reducing costs of searching transaction partners by leveraging internal and external contacts.

All licensing executives of *Alpha Patentfonds* have already worked in different industries. Therefore, *Alpha Patentfonds* is not only able to identify the opportunity to transfer patents to other industries, but it is also able to realize this transfer based on the industries' spanning network.

Search costs are also reduced by transferring this task to patent aggregating companies that are able to reduce the actual number of transactions, as well as the number of unsuccessful approaches to potential buyers.

MPEG acts as a single source for facilitating, organizing, and operating patent pools. A major task of *MPEG* is to offer licenses necessary for a particular technology standard or from multiple patent holders in a single transaction. The producing company that seeks to leverage patents by standardizing technology is able to offer not only own patents but all relevant patents in the technology area without large operative costs. Additionally, multiple patent users are approached by *MPEG* and provided with a multiple license in one step. That reduces search costs and therefore transaction costs.

---

[15] Transaction costs economics states two alternative governance modes to perform transactions: hierarchy and market. Depending on various factors, either market or hierarchy is a better environment to perform transactions. Transactions based on standardized agreements as commodities are better performed in markets. Transactions that need to be controlled for opportunistic behavior are better performed in hierarchy (Williamson, 1975). Transaction costs economics subsequently recognize that hybrid governance modes, as trilateral governance in occasional transactions with high specific goods, are possible (Williamson, 1985).

Based on the experiences of amassing patents, patent aggregating companies are able to realize economies of scale and reduce the costs of negotiating and executing contracts.

> *Acacia* controls over 180 patent portfolios. Based on the experiences of these aggregating activities, the company has developed a semi-standardized due diligence process with selected experts. Additionally, the negotiations follow certain structures and terms, derived from past successful transactions. As a result, the negotiating and the executing process is conducted efficiently.

4.2.3 Potentials for decision making

Patent aggregating companies offer potentials for decision making that affect the company strategy and its innovation strategy. Figure 22 summarizes the potentials for decision making and how they can be realized. Utilizing patent aggregating companies and realizing the potentials for decision making increases the set of strategic choices.

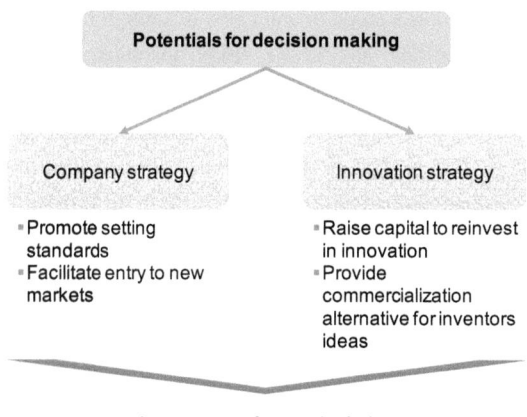

Figure 22: Internal potentials for decision making

Company strategy

Patent aggregating companies offer alternatives to set the general strategic direction of producing companies. Especially in certain industries, the commercialization of a product is only successful if the firm is able to find external adopters of its technology (Conner, 1995; Lichtenthaler, 2005; Reitzig, 2004a). Therefore, companies from certain industries, such as information technology, communication, and chemical industry and medical devices, attempt to establish industry standards of the company's specific technology (Ehrhardt, 2004; Rosenbloom & Cusumano, 1987). If the efforts to establish a standard fail, the company faces severe strategic problems, such as loss of market shares or entire markets.[16] Patent aggregating companies can support companies by establishing standards and help to overcome obstacles that the company would not be able to tackle by itself due to size, resources, or capital constraints.

> *MPEG LA* administers the patent pool MPEG-2. Approximately 1,500 companies have licensed MPEG-2 Patent Portfolio, which includes 880 essential patents in 57 countries owned by 25 patent owners. The MPEG-2 technology is covered with patents owned by many parties. Only when *MPEG LA* offered a viable solution, access to essential patents not accessible for single companies was possible, and the standard was established. Today MPEG-2 is the core technology of most digital television and DVD formats and the most widely employed standard in consumer electronics history.

Historically, out-licensing of patents is used as a mode to enter new or foreign markets (Contractor, 1980). For this, the technology has to be ready to market, and collaboration parties have to be available. SMEs often do not have the resources or the network to establish this cooperation. Assigning patents to a patent aggregating company fosters entry to new markets. The original patent owner can be connected to foreign companies by assigning patents and working closely with the patent aggregating company. This connection can be used for future projects.

---

[16] The race of DVD technology's replacement is an example of a failed standardization attempt and the resulting loses in market share. This race took place between the Blu-ray disc and HD DVD optical disc for storing high definition video and audio. *Hitachi, LG, Panasonic, Pioneer, Philips, Samsung, Sharp, Sony,* and *Thomson* formed the Blu-ray Disc Foundation in May 2002 (Royal Philips Electronics, 2002). *Toshiba, NEC, Sanyo,* and *Memory-Tech Corporation* formed the HD DVD Promotion Group in September 2004 (Toshiba, 2004). Due to essential decisions by major film studios and retail distributors, changing business alliances, and *Sony*'s decision to include a Blu-ray player in the PlayStation 3 video game console, *Toshiba* announced on February 19, 2008, it would cease developing, manufacturing, and marketing HD DVD players and recorders (Toshiba, 2008). The Blu-ray format was established as standard for video and audio players. Analysts estimate the sales volume of Blu-ray players of USD 1.3 billion in 2010 and expect a mass-market penetration and spiking to nearly USD 6.9 billion by 2013 (Gruenwedel, 2011).

Innovation strategy

Patent aggregating companies can extend alternatives for decisions on innovation strategies. Companies that conduct own R&D often have to prioritize research projects due to financial constraint. Even promising research projects have to be terminated because R&D budgets are limited. A smaller number of R&D projects limits the R&D portfolio diversification and increases the risk of the company. Failed projects have a serious negative effect since projects that could compensate the failure are missing. Assigning patents and projects promising, embryonic, or no longer fitting with the company strategy to patent aggregating companies transfers potential future cash flows to the present and increases the budget that can be invested in R&D and the innovation process. Therefore, more innovations can be generated, and more projects can be selected for further development. Additionally, patent aggregating companies provide alternative ways of commercialization and therefore, offer incentives for inventors. The chance to generate actual rent from their innovation, outside the normal product commercialization space, motivates inventors. Motivated employees stay with their company and deliver more innovation. That increases the selection opportunities for later product commercialization.

4.3 Summary

Based on empirical data on patent aggregating companies, as well as derived from literature, this chapter identifies external and internal potentials of patent aggregating companies (Figure 23).

Relating to the general research questions: *Patent aggregating companies as option for producing companies?*, the analysis of the potentials shows that patent aggregating companies can help original patent owners to perceive and to realize patent portfolio leveraging opportunities. Additionally, original patent owners can benefit on a micro level, as patent aggregating companies expand their scope of action.

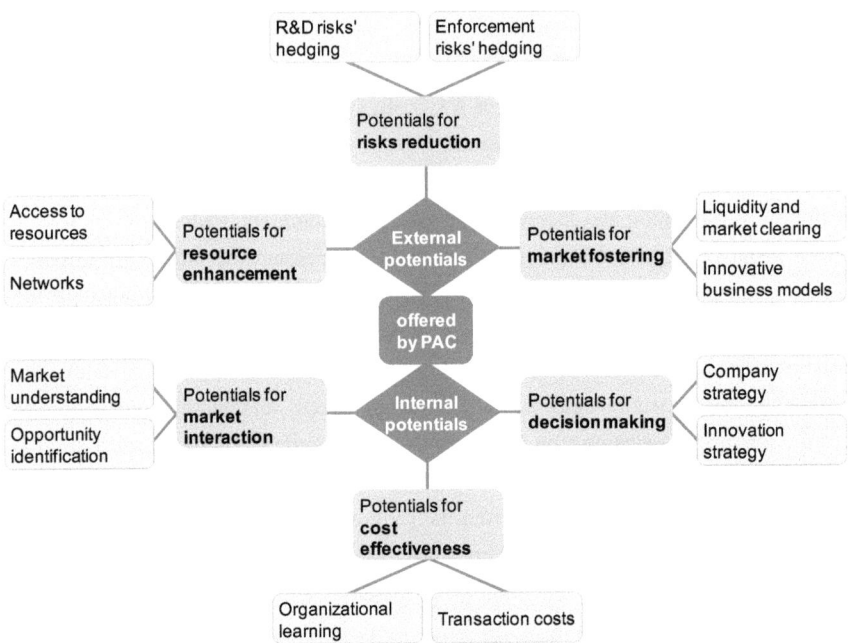

Figure 23: Summary of patent aggregating companies' potentials

As many producing company still face resource constraints and therefore, are often not very successful in the market for patents and technologies, patent aggregating companies can help to overcome these constraints. Patent aggregating companies offer complementary expertise and help to overcome the lack of resources. Within the company, potentials that provide relief from resource constraints can be realized. Collaborating with a patent aggregating company can lead to learning effects, and the original patent owner can start to build up their own resources.

From a macro perspective, patent aggregating companies can foster the development of the market for patents and technologies, which is the basis for all leveraging activities. Therefore, by utilizing patent aggregating companies, the company not only realizes direct potentials, but also paves the way to a more efficient market for patents and technologies.

5 Typology of patent aggregating companies

All patent aggregating companies acquire patents from producing companies and compensate the original patent owners in a certain way. However, this distinction is not sufficient to provide recommendations on how patent aggregating companies can be utilized by producing companies to leverage patent portfolios. Depending on whether the producing company divests only the sole legal right or also plans to transfer technology, different types of patent aggregating companies have to be considered.

In the following chapter, the results from the data analysis of the 27 case companies are used to identify four 'archetypes' of patent aggregating companies. The four archetypes differ significantly regarding their competencies and the rewards they offer the original patent owners. The typology allows helps patent managers of producing companies that wish to optimize their patent leveraging deal with the selection of patent aggregating companies. Therefore, it serves as basis for a management framework developed in section 6.1.

The four archetypes of patent aggregating companies are derived based on three distinctive differences. Two different business models of patent aggregating companies represent each archetype. A case study illustrates the characteristics of each business model and explains the representation of the different archetypes. The illustration of the case study companies is guided by the reference framework developed in section 2.4 and describes the setting, strategy, organization, and the process of patent aggregation of the following eight patent aggregating companies: *Alpha Patentfonds*, *Intellectual Ventures*, *Pete Invest MedTech* [17], *Patent Select*, *Acacia Research Corporation*, *Allied Security Trust*, *MPEG LA*, and *Golden Rice PDP*.

---

[17] The name of the company has been disguised for confidentiality reasons. In this research, the company is referred to under a fictitious name. The name *Pete Invest MedTech* replaces the firm's actual name.

5.1 Four archetypes of patent aggregating companies

Based on their strategies and motives patent aggregating companies have to amass patents, eight types of business models are derived in section 3.3.3: (1) patent acquisition companies; (2) patent enforcement companies; (3) patent incubating funds; (4) patent trading funds; (5) defensive patent aggregators; (6) non-commercial patent aggregators; (7) patent pooling companies; and (8) royalty monetization companies. All eight identified business models aggregate patents from original patent owners and give the owners some benefit in return, but after aggregating the patents, they proceed with the patents in eight different ways.

Analyzing the data and reflecting the needs of the original patent owner regarding their patent portfolio leveraging activities, patent aggregating companies differ not only regarding their motives to amass patents, but also regarding their *competencies* and the *rewards* they offer the original patent owner. Additionally, the empirical findings show that patent aggregating companies differ regarding the *breadth of transaction*.

Competency

Producing companies can leverage their patents by either using the business case or the legal case a patent is related to. Therefore, patent aggregating companies amass patents on one of the two necessary competencies:

i. Business competency.
ii. Nuisance competency.

Business competency. Patent aggregating companies that amass patents based on their business competency have knowledge and detailed information on the underlying technologies. Business competency enables a patent aggregating company to understand the R&D process, the technology, and the product respective business case.

Nuisance competency. Patent aggregating companies that amass patents based on their nuisance competency have legal knowledge and broad experiences in patent monitoring. Nuisance competency enables a patent aggregating company to understand the market for patents, what third parties might do with patent owner's technology, and which legal potential is offered by the infringement case.

Rewards

Producing companies focus on gaining maximal rent from innovation. By utilizing patent aggregating companies for leveraging activities, a producing company can be rewarded in two ways:With monetary short-term rewards that provide producing companies with (additional) cash flows.

i. With monetary and non-monetary long-term rewards that not only include cash flows but also strategic advantages and indirect effects on cash flows.

Monetary short-term rewards. An option to leverage patent portfolios is to focus on generating revenues through selling or out-licensing patents and technologies that do not have any strategic impact on future business. Instead of abandoning these patents, selling or out-licensing them to patent aggregating companies generates lump sum or upfront payments and partial royalties and reduces the costs of the patent portfolio by saving renewal fees. Even if patents still have a strategic impact on future business, producing companies leverage their infringed patents by selling them to patent aggregating companies for the benefit of a lump sum payment over fees received from litigation.

Monetary and non-monetary long-term rewards. Assigning patents to patent aggregating companies does not only have a short-term monetary dimension, but also a strategic dimension. For instance, by donating patents to non-profit organized patent aggregating companies, producing companies can claim tax deductions and create a new marketing tool. Future royalty streams can be transferred to the present and are available for immediate R&D financing. Standards can be created, and new markets are entered based on this standard. Alternatively, learning effects are realized from working together with patent aggregating companies.

Breadth of transaction

Producing companies have to decide whether to leverage their patent portfolios internally, externally, or internally and externally. If they decide to exploit the portfolio externally, another important decision is the breadth of the transaction. Two strategic options are available:

i. Transferring the sole legal right of exclusion (patent).
ii. Transferring the legal right of exclusion in combination with further knowledge and technology.

Sole legal right. Producing companies can sell or out-license the sole legal right of exclusion. In this case, the buyer or licensee only receives the right to use the R&D results but does not receive any further information about the technology or the development process.

Legal right and transfer of technology and knowledge. On the other side, to leverage the patent portfolio optimally, producing companies sell or out-license patents and further knowledge. Drawing from the contributions to knowledge management, knowledge is categorized into information and know-how (Kogut & Zander, 1992). Information means knowledge that can be transferred within or outside firms without loss of integrity once the syntactical rules required for deciphering it are known, for example, blueprints. Know-how comprises accumulated skills and expertise. Therefore, the producing company offers more than the right to use the patent.

Four archetypes of patent aggregating companies

Empirical findings show that patent aggregating companies differ regarding how they support producing companies to leverage their patent portfolios. Patent trading funds, patent acquisition companies, patent incubating funds, and royalty monetization companies are patent aggregating companies that evaluate and exploit patents mainly on the business case and therefore have the internal competencies to understand technologies and develop business cases. In contrast, patent enforcement companies, defensive patent aggregators, non-commercial patent aggregators, and patent pooling companies focus on the legal title and its exploitation and offer competencies to understand other companies' use of patents.

Whereas patent acquisition companies, patent trading funds, patent enforcement companies, and defensive patent aggregators reward original patent owners with monetary short-term rewards, the rewarding of patent incubating funds, royalty monetization companies, non-commercial patent aggregators, and patent pooling companies is based on long-term rewards that would be difficult to achieve for the patent owner without the patent aggregating company.

Four archetypes of patent aggregating companies

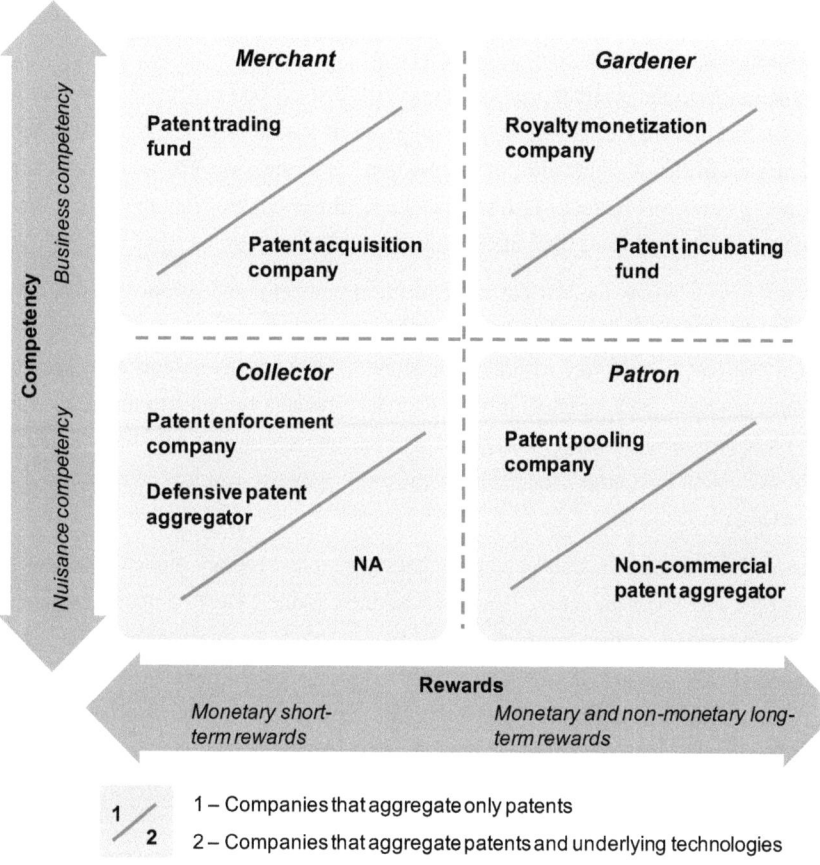

Figure 24: Typology of patent aggregating companies

Based on the two dimensions competency and rewards for patent owners, four archetypes of patent aggregating companies can be differentiated:

i. Merchant
ii. Gardener
iii. Collector
iv. Patron

The competencies and the offered rewards define the four archetypes of patent aggregating companies. Figure 24 illustrates the typology of patent aggregating companies. The vertical axis represents the competencies of patent aggregating companies, and the horizontal axis represents the rewards patent aggregating companies offer to the original patent owners. Two distinct business models represent each archetype but they differ regarding the breath of transaction. The bar within the archetypes indicates the breadth of transaction. Business models under the bar aggregate patents or patents and technologies. Business models above the bar aggregate only the sole legal right of exclusion.

Merchant. The business models patent acquisition company and patent trading fund represent the archetype merchant. The merchant evaluates the patents based on potential markets, as well as potential infringement. They acquire patents and provide short-term rewards. Both business models aggregate patents to generate revenues, but they do it in two different ways. The patent trading fund aggregates only the sole legal right of exclusion. In contrast, the patent acquisition company acquires patents, as well as knowledge or technology.

Gardener. The business models patent incubating fund and royalty monetization fund represent the archetype gardener. These two business models have complete different reasons to aggregate patents but both companies are able to evaluate technologies and market potentials and provide long-term rewards to the patent owners. Through the engagement with the gardener, the producing company is able to foster innovation and finance business growth. Patent incubating funds acquire technologies mainly and use patents as a means of transfer. Royalty monetization companies are interested in existing royalties and therefore, acquire only the sole legal right.

Collector. The business models patent enforcement companies and defensive patent aggregator represent the archetype collector. Resulting from a competency, a monetary reward, and a breadth of transaction point of view, producing companies can utilize the patent enforcement company and the defensive patent aggregator in the same way because both focus on the same patents and compete for infringed patents. The choice of which one is utilized is based on the type itself. Both types help producing companies to prevent being involved in litigation lawsuits while at the same time receiving a certain rent from the used invention. Therefore, both business models have nuisance competency and offer monetary short-term rewards to the original patent owner.

Patron. The business models non-commercial patent aggregator and patent pooling company represent the archetype patron. Both business models solve enforcement issues for the original patent owner and have nuisance competency. The non-commercial patent aggregator, which amasses patents as well as technologies, and the patent pooling company, which focuses only on patents, provide the original patent owners with additional non-monetary rewards and create opportunities for reputation enhancement and gain indirect from R&D.

5.2 Archetype 1 – Merchant

The archetype merchant features business competency and provides patent owners with monetary short-term rewards. As patent trading funds and patent acquisition companies evaluate patents and their business cases and reward patent owners with lump sum payments, these two business models represent the archetype merchant.

5.2.1 Patent trading fund's characteristics

Patent trading funds aggregate patents to generate revenues from acquiring patents or commercialization rights, bundling them to new portfolios, and selling these bundles at a higher price. Patent trading funds buy large amount of patents with a probability in mind that only a certain percentage can be resold. These companies focus solely on reselling the patents. Patent trading funds act as brokers in the market for patents.

Patent trading funds offer investors an opportunity to invest in patents as an asset class. Large financial resources from institutional or private investors back the funding of the aggregating activities, and they operate in a classical investment fund design.

As patent trading funds function in a certain way as brokers, senior management often has patent management experience or a technical background. Additionally, they act as collector and administrator of the invested funds. Several managers also have experience in the financial industry. For legal cases or technical specifications, patent lawyers, patent attorneys, and engineers are employed. The patent trading fund collects investments from private equity, institutional investors, high net worth individuals, or other private investors.

Patent trading funds aggregate patents from all kind of sources. The original patent owners are single inventors, research institutions, SMEs, and MNEs. The original patent owner can recoup R&D investments from abandoned research projects and generate additional short-term cash flows through the actual purchase price, as well as costs saving through transferring the renewal fees to the patent trading fund. Hence, the original patent owner can hedge R&D risks in a certain way.

The patent incubating fund initiates the acquisition process either actively or passively, and it is started before closing the investment fund (so called asset pool[18]) or after closing the investment fund (so called blind pool[19]). Patents are evaluated regarding the criteria fit of targeted technology to the portfolio, application possibilities in other technologies and different markets, market structure and market potential, the legal position of the patent (validity, extent of protection, remaining patent duration), and the anticipated performance and costs of exploitation. After aggregating patents from different companies, the patent trading fund bundles them to new portfolios. Patent trading funds act as the broker and search for the right counterparty to sell the patents. On the one hand, these parties could be from the patent's original industry. On the other hand, patent trading companies are specialized to bundle patents that are then transferable to a completely different industrial sector and could be the entry to new markets or new applications. The contacted companies are corporate buyers of medium or large size, as well as of other patent aggregating companies. In cases of already used patents, the patent trading fund contacts these companies and offers freedom to operate through selling or non-exclusively licenses. In the summary (Figure 25), the cash flows between the involved parties, as well as the transactions of the patents between the original patent owner, the patent trading fund, and the new patent owner respective licensee are illustrated.

---

[18] In an asset pool, the patent portfolio is already known at the time of the investment decision. That means that the investor only carries the exploitation risk (can the patent be exploited successfully and generate return) (Lipfert & Ostler, 2008).

[19] A blind pool first raises money from investors before it invests the raised capital. Therefore, the investors do not know which assets are going to be bought. Blind pools comprise three types of risks for investors: search risk (is the management able to identify valuable patents); purchase risk (is the patent owner of the identified patent willing to sell); and exploitation risk. Due to the higher risks, yield expectations of blind pools are higher compared with asset pools (Lipfert & Ostler, 2008).

Archetype 1 – Merchant

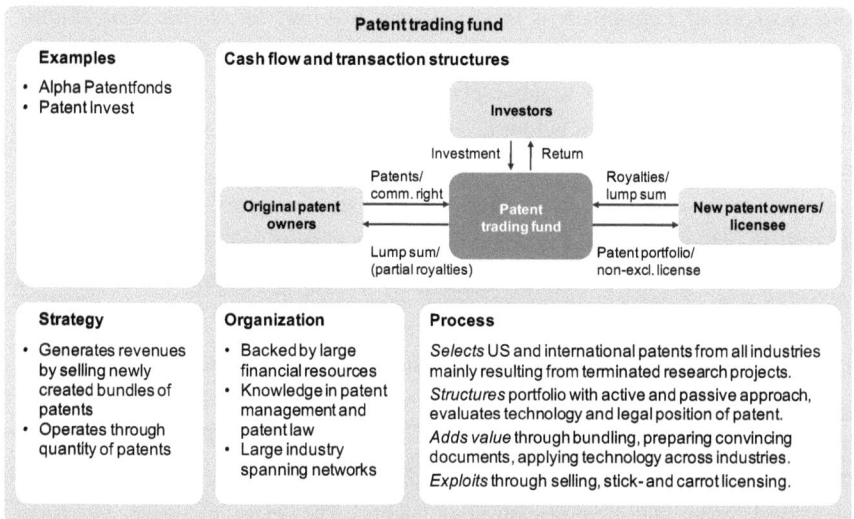

Figure 25: Summary of patent trading funds

The unique characteristic of a patent trading fund is that by aggregating patents, it acts as a match maker between supply and demand, within industries and across industries, backed by funds of private and institutional investors.

5.2.2 Patent trading fund's case study: Alpha Patentfonds

Setting. Alpha Patentfonds is an umbrella term for three investment funds that offer institutional and private investors the opportunity to invest in patents as a new asset class. The three investment funds are Alpha Patentfonds I initiated in 2007, Alpha Patentfonds II initiated in 2008, and Alpha Patentfonds III, where the initiation was split into two tranches in 2008 and 2009. *Alpha Patentfonds* is headquartered in Frankfurt, Germany. The assets under management of the funds differ. Alpha Patentfonds I has assets under management of EUR 32.7 million; Alpha Patentfonds II EUR 49.3 million; Alpha Patentfonds III tranche 2008 EUR 10.3 million; and tranche 2009 EUR 6.23 million. All three funds are closed-end funds[20] and blind pools. The

---

[20] A closed-end fund offers only a limited number of shares and rarely issues new shares after the fund is launched. Shares are only redeemable when the fund liquidates; before this date, the investor has

initiator of *Alpha Patentfonds* is the *European American Investment Bank* (*Euram Bank*) with the head office in Vienna, Austria. *Vevis Gesellschaft für Vermögenswerte* placed *Alpha Patentfonds* in a public placement. The minimum subscription is EUR 10,000. The predicted return before tax is between 17.4% p.a. (Alpha Patentfonds I and II) and 19.6% p.a. (Alpha Patentfonds III Tranche 2009). The planned terms of the funds are four years (Alpha Patentfonds I and II) and five years (Alpha Patentfonds III) with an option for a one-year extension. It was planned that Alpha Patentfonds I ends on March 31, 2011. Due to a change in the market environment, the life of the fund has been extended.[21]

Strategy. *Alpha Patentfonds* aggregates an exclusive right of commercialization from the original patent owners. Therefore, *Alpha Patentfonds* is the exclusive vendor of the patents. The original patent owner still owns the patent but after signing the commercialization contract, *Alpha Patentfonds* is in charge of the exploitation of the patent and owns the commercialization rights. The main objective of *Alpha Patentfonds* is to generate maximum proceeds. The preferred exploitation alternative is the sale of the patents, but if licensing agreements offer higher returns, *Alpha Patentfonds* also out-license patents.

Organization. *Alpha Patentfonds* is the investment company. A Luxembourg-based portfolio company aggregates and exploits the patents. Each investment fund has its own portfolio company. Beside the initiator and the sales partner, *Alpha Patentfonds* has three partners that have been responsible for the selection and the exploitation of the patents. When the funds were set up, *Alpha Gasser Patentverwertungs KG* coordinated the whole process of patent aggregation. *Steinbeis TIB* identifies, evaluates, and selects the patents. In the beginning of 2010, *Alpha Gasser Patentverwertungs KG* merged with *Steinbeis TIB* that now also coordinates the whole process. The original setting planned that *Steinbeis TIB* also exploits the patents but since January 2010, *Charles River Associates* has been in charge of the exploitation of the amassed patents. In some cases, additional consultants are employed for certain

---

to sell the shares at the stock exchange. Additionally, a close-end fund is closed to new capital after it has started operation.

[21] According to Lippert (2011) Alpha Patentfonds has changed its exploitation strategy. Instead of gaining cash inflows through patent sales, Alpha Patentfonds now out-licenses patents and realizes cash inflows over the next 16 years.

stages of the patent aggregating process. For instance, in August 2011, *IP Navigation Group* was retained as a strategic patent advisor to monetize 400 patent assets that the portfolio company of the Alpha Patentfonds II acquired from the *BT Group* (Business Wire, 2011). Figure 26 illustrates the structure and organization of the investment fund Alpha Patentfonds III.

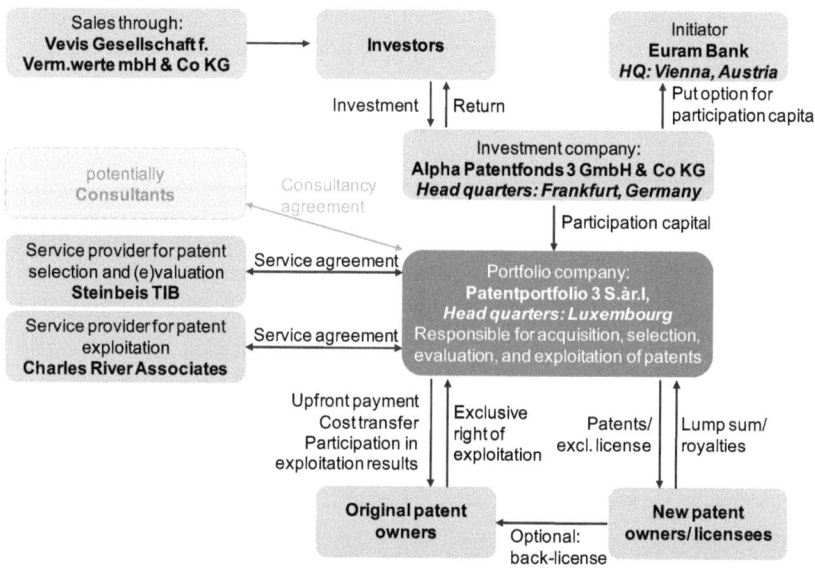

Figure 26: Structure of the organization and relations of participants

Process. Alpha Patentfonds aggregates patents, without any further knowledge or technology, from all technological areas. *Alpha Patentfonds* aggregates only granted patents that are ready for implementation. Only in exceptional cases and as part of a patent portfolio, are patent applications or patents covering embryonic technologies acquired. It focuses on patent owners and companies located in the German speaking part of Europe. Often MNEs are involved and due to international company structures, German, US, and other international patents are aggregated.

Alpha Patentfonds approaches patent owners both actively and passively. When the company started, Alpha Patentfonds approached almost 90% of all patent owners. As a result of the global financial crises and an increasing visibility of *Alpha Patentfonds*, this ratio has changed and the company approaches now only 40% of the patent owners actively. Additionally, patent attorneys connect *Alpha Patentfonds* to promising patent owners.

For structuring the patent portfolio, *Alpha Patentfonds* follows a four-stage process (illustrated in Figure 27). Based on a database of more than 75,000 patent documents covering a large variety of technologies, an algorithm is used to extract patents that fulfill requirements regarding bibliographic data. In addition, patents are consolidated to patent families and patent portfolios. In the second stage, an automatic screening evaluates all remaining ca. 10,000 patent documents regarding the remaining life of the patent, geographic location (at least one national patents has to be granted), proof of concept or prototype, and commercial viability. These criteria are basic indicators for a potentially successful exploitation. As a result, 2,500 out of the 10,000 patents are chosen for further analysis. The patent owners of the 2,500 patents are approached. If they are interested in an exploitation through *Alpha Patentfonds*, they sign a letter of intent that explains their intention to assign the commercialization rights to *Alpha Patentfonds* and the analysis of the patent potential starts. This analysis of potentials evaluates 38 single criteria from five dimensions that are of relevance for external patent exploitation: (1) legal status; (2) market dimension (e.g., commercialization options, technology lifecycle, revenue potentials, opportunities); (3) financial dimension (e.g., production costs, development costs, investments in production); (4) the technology dimension (e.g., unique selling proposition of technology, marketing value, technological advantage); and (5) anonymous interest of potential buyers. Based on the results of the selection process, the patents are aggregated.

After aggregating the patents, an information memorandum is generated for each patent or patent family. This memorandum contains the relevance of the technology for the potential buyer's strategy, the coverage of the technology, potential and existing markets, freedom to operate situation, images, and the contract details. This document is used to approach and convince potential buyers. In addition, business cases are developed to strengthen the selling position.

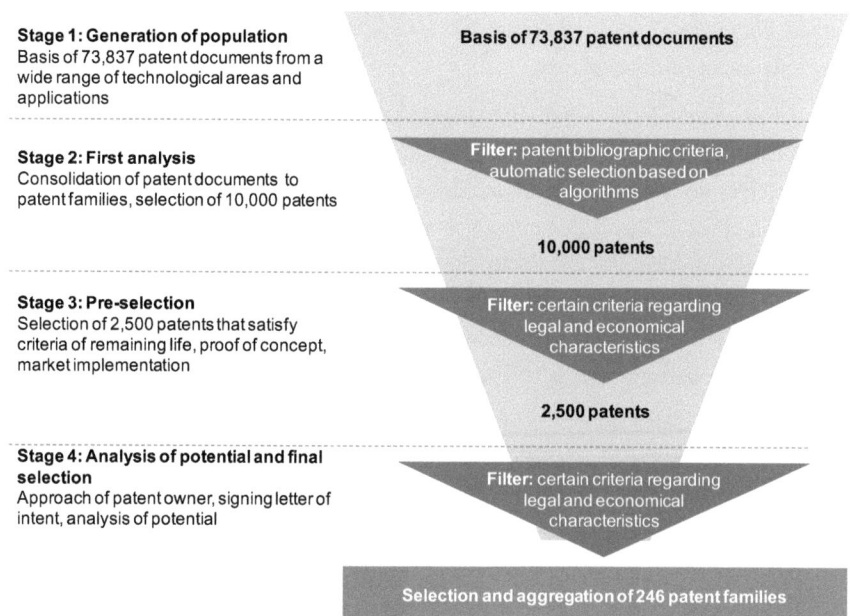

Figure 27: Selection of patents in the structuring phase of the patent aggregating process of Alpha Patentfonds II

The main objective of *Alpha Patentfonds* in the exploitation phase is to generate the maximum proceeds from exploitation. The main channel of patent exploitation is the sale of the patents, but in certain cases, out-licensing is considered. *Alpha Patentfonds* identifies potential buyers based on the existing network and through analysis of patent application data, the market and technology environment, and the potential players in this field. In some cases, patent owners have excluded companies, competitors, or other players from the list of potential buyers. Depending on the technology and the market structure, different strategies of contacting potential buyers are applied. For instance, patents that could be interesting for many companies offer the opportunity either to contact the companies sequentially or to arrange a bidding process. After approaching potentially interested parties, *Alpha Patentfonds* presents the technology and is responsible for negotiations, the signing of the contract, and the post-deal compliance. Additionally, the company supports the buyer in the due diligence process. *Alpha Patentfonds* forecasts that 54% of the aggregated patent portfolio is going to be successfully exploited.

Value for original patent owner. By assigning the commercialization right to *Alpha Patentfonds*, the original patent owner generates an additional cash inflow. The patent owner receives an upfront payment and does not have to cover renewal, valuation, or exploitation costs. If *Alpha Patentfonds* is able to exploit the patent successfully, the patent owner receives an additional 50% of the proceeds. In particular, SMEs, universities, and single inventors often lack the resources, the network, or the competency to exploit their innovation commercially. In collaboration with *Alpha Patentfonds*, the patent owner is able to commercialize the invention in a specific way and recoup R&D investments. MNEs have the competencies to exploit the patents, but often they lack resources to sell, for instance, patents from abandoned research projects. In some cases, back licenses are possible. That prevents the rise of competitors in the patent owner's own application area.

5.2.3 Patent acquisition company's characteristics

Patent acquisition companies aggregate patents to generate revenues from every possible type of patent exploitation. They establish licensing programs, enforce the acquired patents, invest in additional R&D, or apply other exploitation strategies. The basis for this type of business is their ability to benefit from arbitrage. On the one hand, patent acquisition companies are able to detect good patents and buy them at a lower price than it would cost to reinvent these patents. On the other hand, they are able to benefit from the fact that patent portfolios have a greater value than a single patent. Therefore, patent acquisition companies try to increase the value of patent portfolios through the size of it while at the same time, lowering the funding risks.

The business model of patent acquisition companies is based on the quantity of patents they aggregate. Therefore, large financial resources as venture capital or private equity investors, as well as high net worth individuals and institutional investors back patent acquisition companies and their aggregating activities.

The senior management of patent acquisition companies has vast experience in patent management, patent exploitation, or patent law, often from former positions in large industrial companies. A large network in the patent industry is necessary to detect opportunities, and knowledge of patents is important to evaluate opportunities. In specific cases, patent acquisition companies also work with external experts as patent lawyers, financial service providers, or engineers.

Patent acquisition companies amass patents from different industries. Based on a general exploitation strategy, they are able to profit from the different relevance of patents in the different industries. In addition, the transaction breadth is ambiguous. Patent acquisition companies buy patents that are infringed, are close to a technology that is heavily used, or cover a technology that has potential for further development or commercialization. Additionally, they aggregate patents from all relevant geographical markets.

Patent acquisition companies aggregate patents from single inventors, research institutions, and small and large corporate sellers. A producing company can utilize a patent acquisition company to hedge R&D risks. Producing companies can sell patents and technologies of minor or no strategic relevance, which do not fit the company's portfolio any longer, to the patent acquisition company. As a result, the producing company can generate additional short-term cash flows through the actual purchase price, as well as realize cost savings through transferring the liability of the renewal fees to the patent acquisition company. In cases of infringed patents, the original owner is not able or does not want to enforce the patent his/her. For smaller companies not able to commercialize the product or develop the technology further, the option selling the patents to a patent acquisition company generates at least cash flows through the purchase price. In most cases, the original patent owner receives a lump sum payment and does not participate on future revenues.

The patent acquisition company initiates the acquisition process either actively or is approached by the patent owners. After the patent is evaluated regarding its legal position (validity, extent of protection, remaining patent duration), the market structure and the market potential of the targeted product market, existing licensing agreements, pending infringement cases, comparable licensing agreements, and expected performance and costs of exploitations, the patent acquisition company acquires the patent and pays a lump sum to the original patent owner. After acquisition, the main activities of the patent acquisition company are bundling the patents to promising new portfolios or taking advantage of the experience of the employees to exploit undervalued patents. Promising new portfolios could be structured focusing on the opportunity of large infringement lawsuits or to new applications in products. In some cases, the patent acquisition company conducts or mandates other research institutions for further R&D. For the new patent portfolios, the patent acquisition company searches for licensees or buyers. In stick licensing, companies that potentially infringe the patents are detected, and the patent acquisition company offers a license or files a lawsuit. Another opportunity is that already infringed patent portfolios are offered for sale to financial investors or other patent aggregating companies as patent enforcement companies, patent trading funds, or defensive patent aggregators. In carrot licensing or patent sales, companies active in the area of the patent's application are approached and the new created patent portfolios are offered. Additionally, a transfer of knowledge, as blueprints or process know-how, is possible. In the summary in Figure 28, the cash flows between the involved parties, as well as the transactions of the patents between the original patent owner, the patent acquisition company, and the new patent owner respective licensee are illustrated.

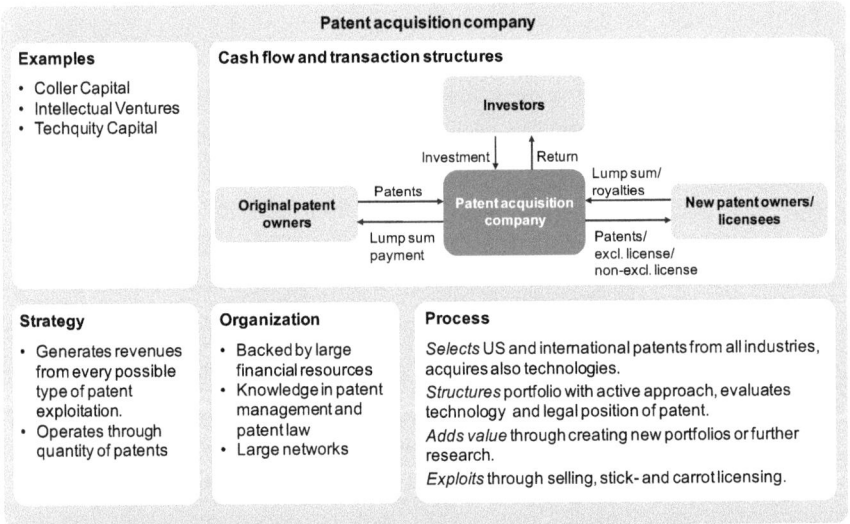

Figure 28: Summary of patent acquisition companies

The unique characteristic of a patent acquisition company is that by aggregating patents, it generates revenues and operates the business based on the mere quantity of its patent portfolio.

5.2.4 Patent acquisition company's case study: Intellectual Ventures

Setting. *Intellectual Venture* was founded in 2000 by Nathan Myhrvold, former Chief Technology Officer of *Microsoft*, Edward Jung, former Chief Software Architect of *Microsoft*, Peter N. Detkin, former Vice President and Assistant General Counsel of *Intel*, and Greg Gorder, former partner at *Perkins Coie LLP*. The patent aggregating company is headquartered in Bellevue, Washington.

To generate capital for its activities, *Intellectual Ventures* has set up four investment funds: Invention Science Fund I, Invention Development Fund I, and Invention Investment Fund I and II. The Invention Science Fund invests in fundamental research and completely new ideas. This fund also finances *Intellectual Ventures'* think tank. The Invention Development Fund invests in the development of already existing ideas that patents do not yet cover. Invention Investment Fund I and II finance the acquisition of patents. Invention Investment Fund I was closed in August 2008 and had a volume of USD 1.5 billion. Sixty to seventy percent of the investors are operating companies; the remaining are from the financial service industry. Invention Investment Fund II has capitalized more than USD 2.5 billion so far. The majority of the investors (ca. 60 to 70%) is from the financial service industry. In total, *Intellectual Ventures* has collected USD 5.5 billion of investor capital.

Intellectual Ventures distinguishes between strategic and financial investors. Strategic investors are mainly operating companies that non-exclusively in-license parts of *Intellectual Venture'* patent portfolios, as well as hold an equity stake in the underlying asset portfolio. Therefore, they have not only financial but also defensive motives to engage with *Intellectual Ventures*. Interestingly, this structure leads to the situation where some companies are investors but at the same time are responsible for a large amount of the return paid to investors. In contrast, financial investors have purely financial motives and, similar to private equity investors receive equity stakes. Investors of *Intellectual Ventures* are technology companies (amongst others *Adobe, Amazon.com, Apple, Google, Microsoft, Nokia, SAP*), university pension funds (amongst others *Brown University, Cornell University, Stanford University, University of Pennsylvania*), individuals, and financial investors and foundations (amongst others *Bush Foundation, Charles River Ventures, Hewlett Foundation, McKinsey and Co., TIFF Private Equity, World Bank*).

Intellectual Ventures owns more than 35,000 US and international patents and patent applications, where ca. 3,000 patents and patent applications are in-house generated or within its inventor network developed. The patent aggregating company has spent more than USD 1.5 billion to acquire patents and patent applications in more than 1,600 acquisition deals. Until now, *Intellectual Ventures* has generated more than USD 2 billion in licensing revenues from ca. 30 licensees.

Strategy. According to official publications, *Intellectual Ventures* sees itself as an invention capitalist and states to invest expertise and capital in the development of inventions. As owner of one of the world's largest and fastest growing patent portfolios it intends to create an active market for invention that connects buyers, sellers, and inventors. Therefore, *Intellectual Ventures* purchases inventions from individual inventors and businesses and combines them into market-specific portfolios, which the company then licenses broadly. Additionally, *Intellectual Ventures* partners with a worldwide network of inventors helping to monetize inventions. *Intellectual Ventures* acquires patents on a large scale. Based on its large portfolio, *Intellectual Ventures* offers licenses or sells the patents.

Organization. *Intellectual Ventures* holds more than 1,100 subsidiaries because for each patent acquisition, a Limited Liability Company (LLC) is formed. Additionally, subsidiaries that represent *Intellectual Ventures* in Asia, Australia, North America, and Europe, are established in eight countries. The company operates and finances its activities mainly through investment funds, which operate in a private equity fashion.

In addition to internal expertise, *Intellectual Ventures* has a wide network of inventors, external engineering, law experts, and freelancers that find patents for and advise *Intellectual Ventures*. For a long time, *Intellectual Ventures* did not litigate patents in its own name but negotiated with potential infringers on the basis of *Intellectual Ventures'* large patent portfolio. A large patent portfolio is expensive to enforce, but for the defendant, it is even more important to defend on the large patent portfolio. Even though *Intellectual Ventures* did not file lawsuits before December 2010 in its own name, several patents previously owned by *Intellectual Ventures* appeared in litigation cases from other companies. Some of these companies were identified as shell companies of *Intellectual Ventures* (Avancept, 2001).

Process. *Intellectual Ventures* acquires patents that cover a broad technology market. The technologies are from agriculture; automotive; communications; computer hardware; construction; consumer electronics; eCommerce; energy; financial services; health technologies; information technology; life sciences; materials science; medical devices; nanotechnology; physical sciences; security; semiconductors; and software. *Intellectual Ventures* aggregates US and international patents, and patent applications with or without the transfer of technology or knowledge.

According to *Intellectual Ventures*, a significant source of patents are single inventors not interested in founding their own businesses, but universities, research institutions, MNEs, and SMEs also sell patents to *Intellectual Ventures*. Additionally, *Intellectual Ventures* focuses on the acquisition of patents from companies in financial distress or that are already insolvent. During the last years, *Intellectual Ventures* has been the biggest buyer in the *Ocean Tomo* patent auctions. In the nine auctions held from April 2006 to March 2009, *Intellectual Ventures* bought 229 of the 302 sold lots.[22]

Besides this traditional ways of buying patents, *Intellectual Ventures* has designed two innovative financial instruments to acquire patents: IP Financing Bridge™ and IP to EPS™. With the construct IP Financing Bridge™, *Intellectual Ventures* offers companies a new source of M&A financing and at the same time, acquires new patent portfolios. Typically, liquid assets, companies' own stocks or debt financing finance M&A deals. If all three financing opportunities are not available or are too expensive for a company that plans to acquire another company, *Intellectual Ventures* provides cash for the M&A transaction. The agreement to provide cash to an acquiring company is based on a contract for assigning the patents of the target company to *Intellectual Ventures*. If the offer for the target company is successful, the acquisition company assigns the target company's patent portfolio to *Intellectual Ventures* and receives a grant-back license. *Intellectual Ventures* benefits from this structure by receiving a large patent portfolio in one transaction. That prevents contacting many single patent owners and offers potentials for transaction cost savings and a lower transaction price.

Offering IP to EPS™ to companies with significant amounts of R&D expenditures and substantial and well-established patent portfolios, *Intellectual Ventures* follows a new way to acquire exclusive rights to sublicense. In the IP to EPS™ arrangement, the patent owner assigns the selected patents to a new subsidiary, which the patent owner

---

[22] *Intellectual Ventures* bought 75.8% of the traded patents at the *Ocean Tomo* auctions. Forty other companies bought the remaining 24.2% (Ewing, 2010).

wholly owns (Figure 29, step 1). The subsidiary, an unrestricted and bankruptcy remote vehicle, grants a free grant-back license. *Intellectual Ventures* acquires an exclusive right to sublicense the patents to any interested licensee from the subsidiary. It pays a fixed guaranteed lump sum and a share of the licensing profits. Additionally, it covers all maintenance and prosecution costs. To reduce further risks, the patent owner receives put rights and claw-back rights. These rights prevent the patent owner from being involved in litigation regarding the assigned patents or allowing them to regain defensive rights when required (Figure 29, step 2).

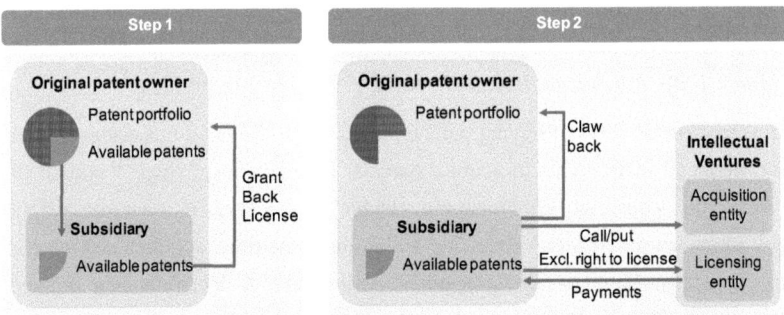

Figure 29: Process of IP to EPS^{TM} transaction

Intellectual Ventures has changed its selection criteria over the years. Based on different dynamics regarding the funding of its two investment funds, Invention Investment Fund I and Invention Investment Fund II, the selection strategy has changed. From 2004 to 2007, *Intellectual Ventures* had a very broad view on the patents it was interested in. The patent aggregating company bought patents from a broad range of industries, along with patents and patent applications from technology lifecycle stages and all commercialization stages, ranging from commercialized products to technologies without any commercial interest yet. The focus of the patent selection was to generate a very large patent portfolio. The average acquisition price for a patent family was USD 35,000. Since 2007, *Intellectual Ventures* has been more selective and focuses mainly on patents with evidence of use or patents already in litigation. Targeted patents have to defend an existing cash flow, be part of an international standard, have to be already in litigation, or have to be claims charts against companies that infringe the patents. The average acquisition price for a patent family is now ca. USD 200,000.

Intellectual Ventures mainly starts the structuring phase of the patent portfolio. The patent aggregating company has subsidiaries in Australia, New Zealand, Canada, China, India, Ireland, Japan, Korea, and Singapore, as well as a large network of freelancers. The freelancers serve as patents scouts either actively finding patents or sent by *Intellectual Ventures* to contact patent owners with interesting patents. They also serve as contact persons for patent owners willing to sell.

After *Intellectual Ventures* has detected an interesting patent, the department 'evaluation services' performs a qualitative evaluation of the patent using a detailed questionnaire. Questions to be answered are: technical quality of invention; legal quality of invention; is there a large addressable market or is it suitable for defensive licensing; is it already involved in litigation (who is involved and how far is the case); has it been licensed (to whom, details); are companies left to license; comparable licensing programs in the industry; is it infringed today (which claims, industry, how large is addressable market, evidence of use); technology lifecycle: technology not commercialized – is it going to be adopted within the next 1 to 5 years (market research, white papers, trends driving technology); legacy technology – how fast the market shrinks; related to a market standard or obligated to a standard setting organization; how are infringements to be detected; is reverse engineering necessary; are there any claim charts; what is a reasonable royalty rate; what are comparables to justify the rates; is the technology core to the business; does the seller claim a grant back license; what is the threshold were the seller to get involved; and the priority date. Currently, *Intellectual Ventures* develops additional analytical tools for patent evaluation to make the evaluation process more transparent and to justify the selection.

The main value adding activity is the bundling of enormous patent portfolios. As *Intellectual Ventures* owns an own research laboratory, in some cases R&D for embryonic technologies is conducted internally or within the network of inventors.

To exploit the patents, *Intellectual Ventures* sells or licenses the acquired patents. The divesture of patents or patent portfolios is done for defensive, as well as offensive reasons. The patent aggregating company also offers a service called IP for Defense (IPfD). This service allows companies that have signed up to the IPfD program to purchase patents from *Intellectual Ventures*. The transferred patents are designated for counter-assertion. Additionally, *Intellectual Ventures* sells patents for offensive reasons. For a long time, *Intellectual Ventures* did not litigate. Therefore, the patent aggregating company sold patents to other companies that enforced the infringed patents. *Intellectual Ventures* main exploitation approach is out-licensing and it offers licensing programs for companies to gain freedom to operate.

Even though founder and CEO Nathan Myhrvold stated, "litigation is a huge failure" and a "disastrous way of monetizing patents" (as cited in Orey & Herbst (2006) in 2006, *Intellectual Ventures* now enforces its patents actively under its own name. In December 2010, *Intellectual Ventures* changed its tactic and filed three patent infringement lawsuits against nine companies. The first defendants were software companies (*Check Point Software Technologies, McAfee, Symantec, Trend Micro*), semiconductor firms (*Altera, Lattice Semiconductor, Microsemi*), and memory chip makers (*Elpida Memory, Hynix Semiconductor*). More lawsuits were filed in July 2011 against *Dell* and *Hewlett Packard*. Therefore, *Intellectual Ventures* now actively follows a stick licensing approach. In becoming a licensee, companies at the same time become strategic investors of *Intellectual Ventures* and hold equity stakes of the patent aggregating company. Until 2009, licensing agreements were mainly closed with MNEs. As licensing activities have started to increase, smaller companies have become involved with *Intellectual Ventures* for amounts in the range of USD 5 million to USD 10 million.

Value for original patent owner. By assigning patents to *Intellectual Ventures*, the original patent owner can generate an additional cash inflow. In general, the patent owner receives a lump sum payment and does not have to cover renewal, valuation, or exploitation costs. Sometimes, *Intellectual Venture* and the original patent owner agree on profit sharing. In this case, the original patent owner receives only a small upfront payment and is rewarded with a profit-sharing back-end. SMEs, universities, and single inventors often lack the resources, the network, or the competency to exploit their innovation commercially. In assigning the patents to *Intellectual Ventures*, the patent owner is able to commercialize the invention in a specific way and recoup R&D investments. In addition, in stages where the invention is not yet ready for the market, the patent owner does not have to take the risk of further, potentially fruitless, development but can still benefit from the invention. MNEs have the competencies to exploit the patents themselves but often these companies lack resources to sell for instance, patents from abandoned research projects.

Additionally to traditional selling transactions, patent owners can benefit from *Intellectual Ventures'* financial products IP Financing Bridge™ and IP to EPS™. For companies in an acquisition process, IP Financing Bridge™ serves as a bridge loan to finance the acquisition if other financial assets are not available or too expensive. In case of a successful transaction, the bridge loan does not have to be repaid. Instead, *Intellectual Ventures* gains ownership of the target company's patent portfolio, and the company that acquires another company receives a grant back license.

Using IP to EPS™ instead of selling the patents, the original patent owner can leverage R&D risks and smooth volatile rents from its invention while still able to exploit its invention. An additional benefit is that normally a patent sale or a onetime patent settlement is classified as other income and does not affect earnings on the balance sheet. Structuring the deal correctly, *Intellectual Ventures'* cash payment to the patent owner is treated as earnings. The patent owner also saves costs by shifting maintenance and prosecution costs to *Intellectual Ventures*.

5.3 Archetype 2 – Gardener

The archetype gardener features business competency and provides patent owners with monetary and non-monetary long-term rewards. As royalty monetization companies and patent incubating funds evaluate patents and their business cases and reward patent owners not only with lump sum payments but also with continuous payments to improve the financial situation, organizational learning opportunities, the transfer from commercialization risks, and insurance against losses of future cash flows, these two business models represent the archetype gardener.

5.3.1 *Royalty monetization company's characteristics*

Royalty monetization companies aggregate patents as security for the capital they provide to patent owners. They collect funds of private and institutional investors to pass to capital seeking companies, and the patents serve as security and are used to ensure that the investors regain at least part of their invested money.

Royalty monetization companies are investment companies that bring together investors from capital markets with companies looking for alternative sources of capital. To provide companies with capital, the patent aggregating company either purchases certain royalty streams or lends money based on royalty streams. In a royalty purchase transaction, the capital seeking company receives an upfront payment and assigns all or a portion of their future royalty inflows to the royalty monetization company. In general, the original patent and royalty owner does not have to pay the money back but at the same time, they do not have any rights on the remaining assigned royalties. If the royalty monetization company lends money to the capital seeking company, it retains residual ownership of the royalties once the bond is repaid. In many of the cases, the provided upfront payment is structured as a royalty bond. In a royalty bond, royalty interests are bundled, securitized, and sold at the capital market. As security, the royalty monetization company acquires the patents and concomitant licenses through a special purpose vehicle (SPV) in a true sale transaction.

Senior managers of royalty monetization companies have a financial or scientific background. In royalty monetization transactions, the product that the patent covers and the resulting cash flows have to be evaluated rather than the patent itself. For a legal assessment of the patents, royalty monetization companies employ external resources, such as a patent attorney or a patent valuation service firm. Additionally, legal and financial advisors are employed to design the securities to be sold.

Royalty monetization companies focus mainly on patents from the pharmaceutical industry, ones that are already licensed to third parties and cover products approved, or in stage III of the approval process by the FDA. For the investment in royalties, steady and long-term cash flows are necessary. This requirement is difficult to meet in other industries. Patents serve only as security, and royalty monetization companies aggregate only the legal right of exclusion without any underlying knowledge.

Royalty monetization companies are mainly engaged with research institutions or small and medium biotechnology or biopharmaceutical companies. Royalty monetization companies take over the risk from out-licensing the R&D results and insure the original patent owners against a loss of royalty payments. Hence, a producing company can utilize a royalty monetization company to hedge R&D risks. Instead of waiting for the royalties and reinvesting them annually, the original patent owners receive the equivalent immediately and can reinvest it in new R&D projects, marketing activities, or other company related activities and create a long-term advantage. Selling the royalties and patents to the royalty monetization company is also insurance for the original patent owner. The lump sum payment the patent owner receives is a form of non-recourse debt. If the royalty payments stop unforeseen, the royalty monetization company has to deal with the loss of cash flows. The original patent owner has received the originally expected amount of money.

Basis for the work of royalty monetization companies are existing licensing agreements between the original patent owner (licensor) and another company (licensee) that generate predictable cash flows. The licensing agreements are closed without the involvement of the royalty monetization company. In general, patent owners approach royalty monetization companies and offer royalty streams and patents after the licensing deal is closed. In a royalty purchase transaction, royalty monetization firms invest from already collected blind pools. In royalty bond transactions, the royalty monetization company designs a bond-like financial instrument. Buying the bond at the capital market, investors know which royalties and patents are the underlying for the bond. The SPV issues bonds that raise the patent's purchase price paid to the original owner. The royalty interests from the license of the patent back the bonds. After transferring the patents to the SPV, the licensee pays the royalties not to the licensor but to the SPV. In the summary (Figure 30) the cash flows between the involved parties and the transfer of the patents or licenses between the original patent owner and the SPV are illustrated.

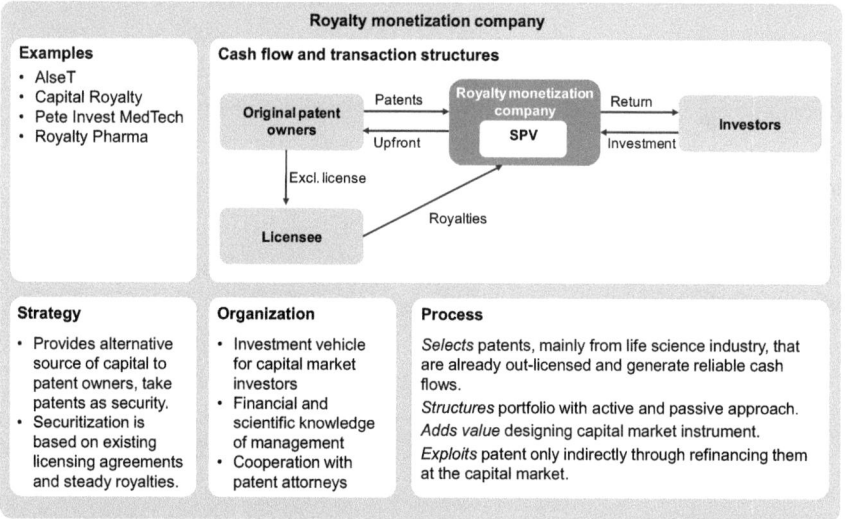

Figure 30: Summary of royalty monetization companies

The unique characteristic of a royalty monetization company is that by aggregating patents, it shifts future royalties to the present and provides the original patent owner with immediate capital and an insurance against loss of royalties.

5.3.2 Royalty monetization company's case study: Pete Invest MedTech

Setting. *Pete Invest MedTech* [23] is a division of *Pete Invest Partners*. *Pete Invest Partners* is US-based private equity company that focuses on secondary private equity and secondary venture capital investments. The founder founded it in the 1991 when he had the opportunity to acquire a significant number of venture capital and leveraged buyout fund positions. The company extended its business and founded *Pete Invest MedTech* in 1999 to offer a platform for investors to invest in royalty and revenue interests of drug and medical device products.

Until now, *Pete Invest MedTech* has established three investment funds: Pete Invest Sector Funds I (asset under management ca. USD 300 million), Pete Invest Sector Funds II (asset under management ca. USD 600 million), and Pete Invest Sector Funds III (asset under management more than USD 1 billion). The funds invest in royalties and revenues streams of healthcare products. The number of investments *Pete Invest MedTech* has made varies between the funds. The first fund financed ca. 20 investments with an average deal size of USD 15 million. The second fund focused on larger investments and financed ca. 19 investments with an average size of USD 45 million. The third fund is still being invested and the size of the investment varies. It plans to invest in ca. 40 deals, with an average investment of ca. USD 30 million. Main investors of the funds are pension funds, high net worth individuals, and family offices.

With these funds, *Pete Invest MedTech* focuses mainly on purchasing royalty streams. The company has only been involved in three royalty bonds. One of the biggest deals with large publicity was the issuance of more than USD 200 million of asset-backed notes. A portfolio of healthcare royalty interests and revenue interests in more than 20 biopharmaceutical products, medical devices, and diagnostics selected from two thirds investments made by Pete Invest Sector Funds II backed the notes.

---

[23] The name of the company has been disguised for confidentiality reasons. In this research, the company is referred to under a fictitious name. The name *Pete Invest MedTech* replaces the firm's actual name.

Strategy. Pete Invest MedTech is an investment company that offers financial investors the opportunity to participate in the success of life science products that are covered by patents and are uncorrelated to other assets. To generate an optimal rate of return, *Pete Invest MedTech* purchases royalty interests or revenue interest from promising life science products that have already completed all of the development activities. Early stage products are only considered when they are in a pool with approved products. Patents are aggregated to serve as security. *Pete Invest MedTech* does not conduct further patent management or patent exploitation.

Organization. Pete Invest MedTech's experiences are in the area of healthcare (including clinical research, sales, and marketing operations) and finance (structured finance, venture capital, investment banking, and capital markets). The main objective of the company is to structure transactions that suit capital seeking transaction partners, as well as return demanding investors. Therefore, internal competencies are the evaluation of the potential of healthcare products and transaction structuring.

Pete Invest MedTech operates with a network of experts, and it commissions several service providers for each deal. Placement agents, for instance, raise the capital for the funds. They contact potential investors, introduce *Pete Invest MedTech* to large institutional investors, set up introductory meetings, and help to craft an offering memorandum and prepare the pitch.

Process. Pete Invest MedTech focuses on the acquisition of royalty streams of life science products that have completed all development activities and are ready to or already commercialize. Therefore, the company diversifies its investments across products, specialties, clinical and regulatory stages of development, and geographic markets. Only the patent without any underlying technology or knowledge is transferred to a SPV held by *Pete Invest MedTech*. Patent owners whose royalties are monetized by *Pete Invest MedTech* are small and medium biotechnology companies, academic and research institutions, and big pharmaceutical companies. Most often, *Pete Invest MedTech* acquires royalty streams from medium sized companies, with USD 10 million to USD 100 million sales per year.

Pete Invest MedTech selects its investments depending on the potential of the life science product. Investments in royalties are passive investments without any influence on the outcome after the actual investment. *Pete Invest MedTech* acquires revenue streams from its transaction partner but these revenue streams do not result from the ultimate product but are paid by a third company that has licensed the product. That creates additional risks for *Pete Invest MedTech,* hence the structuring phase and the due diligence process are essential steps in the patent aggregating process. In general, *Pete Invest MedTech* follows a passive acquisition approach, and patent owners who want to monetize royalties approach the company. Only approved products that are already licensed enter further evaluation and due diligence process. After this initial check, *Pete Invest MedTech* focuses on an analysis of the underlying patents. The company commissions a patent law firm to evaluate the types of patents involved in the product, the strength of the patent portfolio, if the patent portfolio comprises the product, and the opportunities to circumvent. The results show the strength of the patents at stake. Based on the results, *Pete Invest MedTech* analyzes other players and patents associated with the respective patent and determines the strength, quality, breadth, and the term of the patent. If the patent evaluation is finished with a positive result, *Pete Invest MedTech* evaluates the commercial potentials and the licensee of the product internally. After developing a forecast model and talking to a number of key opinion leaders, a third party, which can be a consulting firm, an individual consultant, or another specific expert, is employed to build their own forecast model and challenge *Pete Invest MedTech's* assumptions. In cases of a positive due diligence process and an investment of *Pete Invest MedTech,* the underlying patents are transferred to a SPV.

Due to the passive nature of the investment, *Pete Invest MedTech* is not involved in any value adding activities and only rarely in the exploitation phase. If the patent is transferred to the SPV, it only serves as security, and further value adding activities are not conducted. Only in the case of bankruptcy of the original patent owner would *Pete Invest MedTech* attempt to sell or out-license the patent to reduce loss in the payback cash flows. However, this has not happened yet.

Value for original patent owner. By assigning patents and transferring royalty interest to *Pete Invest MedTech,* the patent owner can transfer commercialization risks to the patent aggregating company. Transferring the royalty streams to *Pete Invest MedTech* also reduces the financial risk of loss of royalties in the future. Additionally, companies that are engaged with *Pete Invest MedTech* have a capital need. Life science products often create large royalty streams over a considerable time period. Transferring the future royalty streams to the present can satisfy the urgent demand for capital. Advantage of royalty monetization is its non-dilutive nature and the fact that it is also available for companies that cannot tap the debt market. Receiving immediate capital companies can invest in R&D, marketing, or sales activities.

5.3.3 Patent incubating fund's characteristics

Patent incubating funds aggregate patents to exploit the underlying technology and to generate revenues from a carrot licensing approach. They aggregate patents, invest in further R&D, and out-license the enhanced technology to other companies.

Patent incubating funds offer investors an opportunity to invest in patents as asset class. Therefore, large financial resources from institutional or private investors back the funding of the aggregating activities and they operate in a classical investment fund design. The difference between patent incubating funds and patent trading funds is that by investing in patent incubating funds, investors participate in value creation, whereas by investing in patent trading funds, investors participate in arbitrage.

Patent incubating funds operate mainly as collector and administrator of the invested funds. Therefore, the senior management has a general management background or experience in the financial industry. To select, advance, and exploit the patent portfolios, patent incubating funds use a large network of engineers, patent service providers, research institutes, patent attorneys, patent lawyers, and patent intermediaries.

The quality of the aggregated patents is as important as the commercialization opportunities of the technology. Therefore, patent incubating funds focus mainly on promising, often embryonic, technologies from a broad range of industries that have the potential of successful commercialization in products. Not only is the legal right transferred but so is additional knowledge.

Patent incubating funds aggregate patents mainly from single inventors, research institutions, universities, and SMEs. A producing company can utilize a patent incubating fund to hedge R&D risks. Often the original patent owner does not have the financial resources to develop the technology further and to commercialize it. Patent incubating funds also acquire the terminated research projects of MNEs with promising technology and high quality patents. In selling these patents and technologies to a patent incubating fund, the producing company can generate cash flows through the actual purchase price, and realize costs saving through transferring renewal fees and further R&D costs to the patent incubating fund. In addition, options exist for a back license and long-term monetary rewards through participating in the royalties of the commercialized technology.

The patent incubating fund collects investments from private equity, institutional investors, high net worth individuals, or private investors. The acquisition process is initiated either actively or passively by the patent incubating fund and it is started before or after closing the investment fund. The collected funds are invested firstly in the aggregation of patents. For this, the evaluation of the patents focuses especially on the quality and future prospects of the underlying technology. Determinates of the evaluation process are the commercialization potential of the technology, the expected time to market, the market structure and the market potential of the targeted product market, the legal position of the patent (validity, extent of protection, remaining patent duration), and the anticipated performance and costs of exploitations. After aggregating the patents, the patent incubating fund mandates external R&D institutes for the advancement of technologies and patents. In the advancement phase, technologies are improved or scaled up, or industrial proof-of-concepts are achieved. Improvements, as well as circumvent solutions, are patented continuously, and the newly created patents improve the strength and the scope of the patent portfolio. Additionally, the patent portfolio is advanced to protect the technology in international markets. The collected investment fund also finances further development and advancement, and includes, for instance, prototyping or expanding the geographical scope of the patents. After the advancement phase, the fund follows a carrot licensing approach and searches interested companies. These companies are either large companies filling their product pipeline with the offered technology, or smaller companies benefiting from the new technology by diversifying their product portfolio without being involved in the risk of R&D. In most of these deals, a transfer of knowledge or technology is involved. The patent incubating fund's objective is to sell or exclusively out-license the advanced technology to a sharply higher price and repay investors, and after deducting additional R&D costs and administration fees, this is at a higher rate of return than the return of traditional investment funds. In the summary (Figure 31), the cash flows between the involved parties and the transactions of the patents between the original patent owner, the patent incubating fund, and the new patent owner respective licensee are illustrated. Additionally, the involvement of the mandated R&D institution is illustrated.

Archetype 2 – Gardener

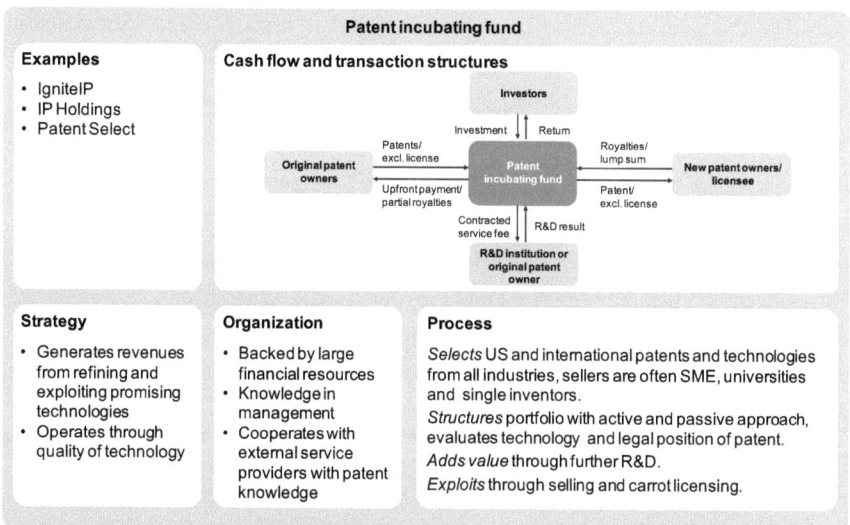

Figure 31: Summary of patent incubating funds

The unique characteristic of a patent incubating fund is that by aggregating patents, it is able to incubate embryonic technologies through the employment of a large network of service providers based on funding of private and institutional investors.

5.3.4 Patent incubating fund's case study: Patent Select

Setting. Patent Select is the umbrella term for three investment funds that offer private investors the opportunity to invest in patents as a new asset class. *Deutsche Bank* and *Clou Partners* initiated all three investment funds. In 2006, Patent Select I was initiated, Patent Select II and Patent Portfolio I followed in 2007. The asset under management of the funds differs between EUR 24.7 million (Patent Select I) and EUR 130 million (Patent Portfolio I). All three funds are closed end funds. Patent Select I and II are constructed as asset pools; Patent Portfolio I is a partly blind pool. *Deutsche Bank* sold Patent Select I and II as public placements. The minimum subscription for the two Patent Select funds was EUR 50,000. They were offered to private clients. The subscription capital of the public placed Patent Portfolio I was EUR 10,000. The predicted return before tax is ca. 12%. The planned terms of the funds are six to eight years. The funds invest ca. 5 to 7% of the shareholders' equity to aggregate patents. The two *Patent Select* funds each aggregated 12 patents respective patent families, and Patent Portfolio I aggregated 22. The largest portion of the total shareholders' equity is used to finance the development phase.

Strategy. Patent Select acquires embryonic technologies not ready for product commercialization and the patents that cover these technologies. Patent owners transfer the patent rights to *Patent Select* and give up their ownership rights but are still involved contractually for providing their knowledge in the further development. After transferring the patents, *Patent Select* starts with further development, additional R&D, and other advancement measures to develop a product ready for market. After concluding this, *Patent Select* exploits the patents, depending on the application and the market, through carrot licensing or sales.

Organization. The main involved parties in *Patent Select* are *Deutsche Bank, Clou Partners*, and, until August 2010, *IP Bewertungs AG. Deutsche Bank*, the largest German financial institution, together with *Clou Partners*, is the initiator of the funds. As initiator, they are in charge of the set up of the funds and in compiling the prospectus of the investment opportunity. Figure 32 shows the structure of the investment fund Patent Portfolio I. *Neunzehnte Paxas* and *ZEA Beteilungsgesellschaft* are in charge of the fund management. At the time of initiating the funds, *IP Bewertungs AG* was mandated as service provider. The service provider is in charge of the identification, selection, evaluation, allocation, and management of the patents. In July 2010, *IP Bewertungs AG* filed for insolvency and was replaced with new service providers for the exploitation of patents.

Archetype 2 – Gardener

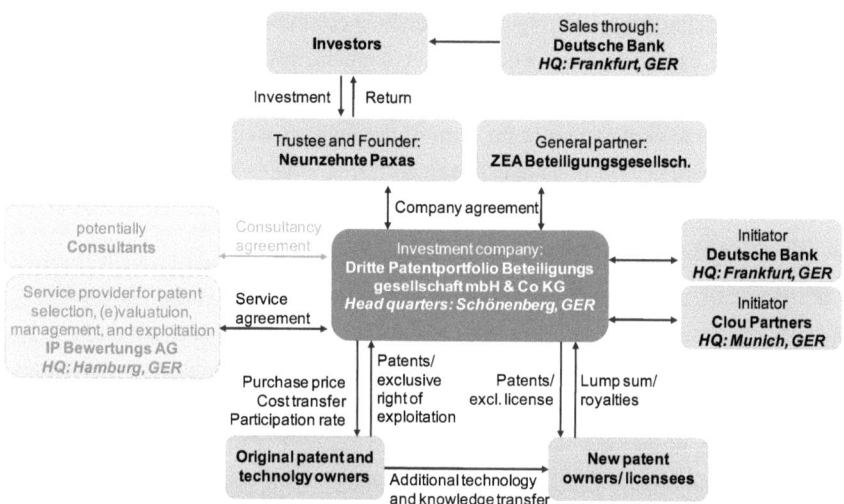

Figure 32: Exemplary structure of the organization and relations of participants illustrated on the investment fund Patent Portfolio I

Process. Patent Select aggregates embryonic or still unmarketable technology and patents that cover these technologies from all technological areas except those covering military or genetic engineering. These patents do not fit the 'Reputational Risk and Corporate Governance Criterias' of initiator *Deutsche Bank*. *Patent Select* also evaluates patents from all geographic areas but due to the strength of the German Patent Office, many aggregated patents are German. Even that *Patent Select* evaluates patents from all kind of companies and inventors, universities or research institutions and SMEs develop most aggregated inventions.

Based on a large network of service providers and presence on global technologies fairs, the patent owners are mainly actively approached. *Patent Select* decomposes the evaluation process in a quantitative pre-selection stage and a qualitative audit of interesting patents (see Figure 21 for the different stages of the evaluation process). After passing the pre-selection, the qualitative audit starts. The technologies and patents are evaluated regarding 'asset criteria' and 'fund's criteria'. Asset criteria contain (1) technical criteria (e.g., considering the market standard, examining the technology lifecycle and existing trends, is time to market shorter than 36 months); (2) legal criteria (e.g., attribute analysis, status, scope of patent, freedom to operate, potential for circumvention, infringements); and (3) economic criteria (e.g., evaluating the market and exploitation potential, market segment, market volume). The fund's criteria focus on the sufficiency of patents for an investment fund and evaluate the fit of expected time to exploitation to the term of the fund, as well as the expected R&D costs to reach exploitation potential. Along with this, *Patent Select* drafts a strategy plan for commercialization. Based on the results of the evaluation and the strategy plan, *Patent Select* acquires the patents.

In the phase of value-adding activities, which is two to four years, *Patent Select* mandates, for instance, R&D institutions to develop technologies that are ready to be commercialized in products. For instance, research laboratories enhance technologies from the life science industry to manufacture marketable products. Engineering technologies are further developed, and proof of concept is delivered by producing prototypes. Additionally, exhibitions and technology fairs promote the inventions. During this phase, the original patent owner is still involved. The original patent owner conducts some of the development activities or is involved through a consultancy contract.

Patent Select often approaches potential licensees or patent buyer before the advancement of patents starts. With this strategy, *Patent Select* is able to offer customized solutions, for example, prototypes or adjustments regarding certain environmental influences or already existing production facilities, to interested parties and can increase the success rate of exploitation and the exploitation return. The patents are exploited through carrot licensing or sales, depending on the market structure, the number of potentially interested parties, and scope of application. For instance, exploiting a technology that dispenses liquid in a new way and has 40 different applications an approach offering exclusive licenses for different regions or applications would be chosen. For a patent that covers a technology for only few applications in a monopolistic market, the patent is sold to the only interested party. Due to the strict selection process and the focus on few patents of high quality, *Patent Select* forecasts that all aggregated patents and patent families are going to be exploited successfully.

Value for original patent owner. By assigning patents to *Patent Select*, the original patent owner can generate an immediate additional cash inflow and future participation rates. In particular, SMEs, universities, and single inventors often lack the resources, the network, or the competencies to develop their invention further and present proof of concepts or prototypes to potentially interested parties. Assigning the patents to *Patent Select*, the invention is further developed and commercialized in products, and the patent owner is able to recoup R&D investments. Also, in stages where the invention is not yet ready for the market, the patent owner does not have to take the financial risk of further, potentially fruitless, development but can still benefit financially from the invention. Often the original patent owner conducts the advancement phase. Therefore, the original owner not only profits financially but also through organizational learning and adaption of competencies without carrying the financial risk of the learning and the R&D processes. Being closely involved in the development process, patent owners are able to extend their network and use the licensing partners for further cooperation, for example, internationalization of future products. The patent owner receives a relatively, low upfront payment and does not have to cover renewal, valuation, or exploitation costs. *Patent Select* acquires patents to moderate prices because after exploiting the technology through *Patent Select*, the original patent owner receives partial proceeds. This payment scheme also sets incentives for the original patent owner to stay involved in further development.

5.4 Archetype 3 – Collector

The archetype collector features nuisance competency and provides patent owners with monetary short-term rewards. As patent enforcement companies and defensive patent aggregators detect infringements, base their work on the legal case of patents, and reward patent owners with a lump sum payments, these two business models represent the archetype collector.

5.4.1 Patent enforcement company's characteristics

Patent enforcement companies aggregate patents to generate revenues from a stick licensing approach. They enforce the aggregated patents and establish licensing programs.

Patent enforcement companies are the pioneering category of patent aggregating companies and have emerged at an early development stage of the patent aggregating business. Patent enforcement companies developed from producing or service companies, from a single inventor, or based on entrepreneurial spirit. Due to the heterogeneity of business history, patent enforcement companies have several sources of funding. They can be public companies, privately owned but also backed by large financial investors, depending on the initial point of business and the initial objectives. Furthermore, the funding of the patent transactions varies. While some patent enforcement companies use only internally generated financial resources for patent aggregating activities, others are provided with investments of private equity firms to finance patent aggregating activities.

Based on the different business histories, the professional background of senior management is diverse and can span from engineers that were inventors in the beginning, to patent professionals, to hedge fund managers or investment bankers. Internal resources are often limited and mainly focus on strong legal and licensing knowledge. Patent enforcement companies rely on a large network of external resources, such as patent lawyers and engineers that they mandate for specific cases.

Patent enforcement companies focus mainly on patents covering technology in the high-technology industry, such as semiconductor, software, information and communication technology, wireless, consumer electronics, but are not limited to these technologies. However, the industry is only of lower priority. Patent enforcement companies focus on patents already in use and possibly infringed covering products in large markets. Patent enforcement companies aggregate only the legal right of exclusion without any underlying knowledge or technology. The aggregated patents often have a broad scope and overlap with other patents. Until now, patent enforcement companies have aggregated mainly US patents, but they increasingly target German, French, or UK patents.

Patent enforcement companies aggregate patents from single inventors, research institutions, and small and large corporate sellers and take over the enforcement risks from producing companies. In case of single inventors, research institutions and SMEs, the original owner does not have the financial resources to enforce the patents. In case of MNEs, the original patent owners often do not want to be involved in patent infringement suits due to strategic or reputational reasons. For instance, in oligopolistic markets, a patent enforcement lawsuit could start a chain of reactions where all players sue each other. To prevent this, the MNE sells the potentially infringed patent to a patent enforcement company. Another strategic reason is risk diversification. Litigation lawsuits are expensive and the outcome is often uncertain. By selling infringed patents, the producing company can generate additional short-term cash flows through the actual purchase price. If the original patent owner is not involved in the enforcement activities, they receive a lump sum payment and do not participate on the generated revenues. In cases where the patent owner is involved in the enforcement activities, the generated licensing revenues are split between the original patent owner and the patent enforcement company. Additionally, the producing company is able to save the costs of litigation.

The acquisition process is initiated either actively or passively by the patent enforcement company. After the patent is evaluated regarding its legal position (e.g., validity, remaining patent duration), potential infringements, comparable licensing agreements, and expected performance and costs of a litigation lawsuit, the infringed patents or their exclusive licenses are assigned to a SPV owned by the patent enforcement company. In some cases, patent enforcement companies create new patent portfolios that cover a certain technology and bundle patents from various patent owners; in other cases, the patent portfolio contains only one patent family. The fund then contacts potential users of the patents either by letter and negotiation or by filing a lawsuit immediately. If the targeted companies, which are producing companies of all sizes, in fact use the patent, they are forced to take a non-exclusive license from the patent enforcement company. In the summary (Figure 33), the cash flows between the involved parties and the transfer of the patents or licenses between the original patent owner, the patent enforcement company, and the licensees are illustrated.

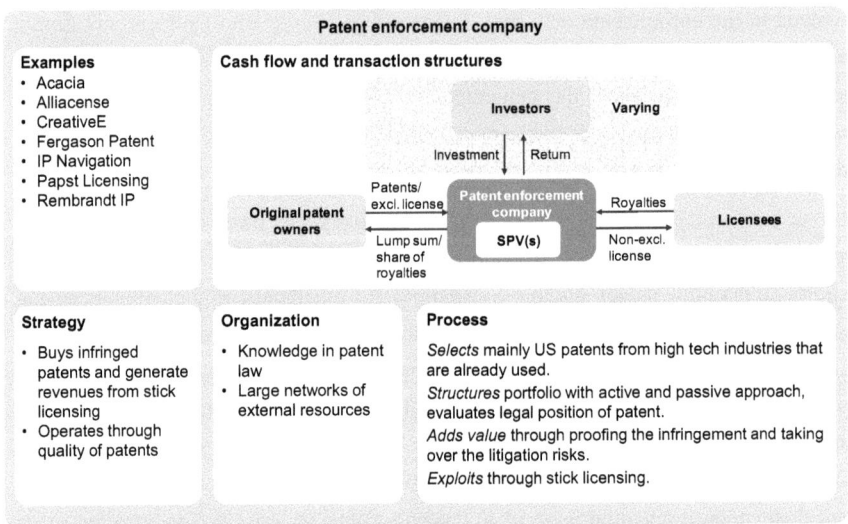

Figure 33: Summary of patent enforcement companies

The unique characteristic of a patent enforcement company is that by aggregating patents, it exploits aggressively already infringed patents with legal means.

5.4.2 Patent enforcement company's case study: Acacia Research

Setting. Starting as a venture capital company in 1993, *Acacia Research Corporation*, based in Newport Beach, California, is today a large and well-known patent aggregating company. After the burst of the dotcom bubble and being stranded with patents and technologies from venture capital investments, *Acacia* started its patent aggregating and patent licensing business in 2003. In the same year, *Acacia* went public. The company is listed at the NASDAQ with the ticker symbol ACTG. Analyst coverage is provided, for instance, by *Barclays Capital* (Darrin D. Peller), *J.P. Morgen* (Paul Heller), or *GARP Research & Securities* (George Sakellaris).

Acacia controls 536 US patents in 332 patent families. These patents are pooled in more than 180 patent portfolios. The patents cover technologies used in a wide variety of industries, for instance, communication technology, consumer electronics, database technology, or software. Between 2005 and 2010, *Acacia's* patent portfolio grew by ca. 430%, and between 2008 and 2010 by ca. 120%. Within the last 10 years, *Acacia* has completed more than 1,000 licensing agreements covering 99 technologies. *Acacia's* sales have increased significantly since going public and in 2010, the company generated a profit for the first time. Figure 34 illustrates the sales and the EBIT of Acacia since its initial public offering.

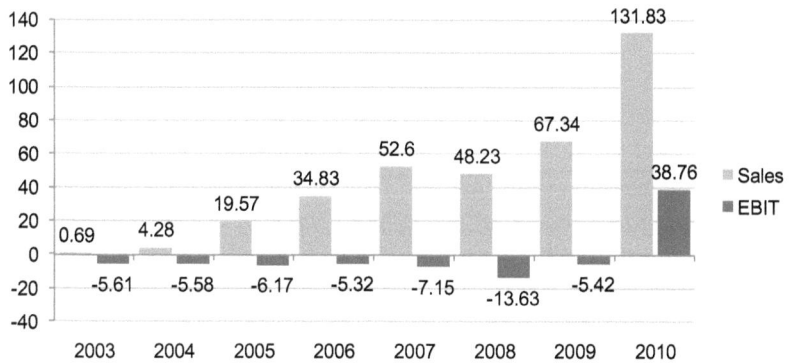

Source: Thomson One Banker retrieved on September 6, 2011

Figure 34: Sales and EBIT of Acacia from 2003 to 2010, in million USD

Acacia is ranked amongst the five most litigious patent aggregating companies. According to PatentFreedom (2011a), *Acacia* has been involved in 384 cases with 1134 counterparties in US courts until the end of 2010.

Strategy. *Acacia's* mission states that the company

> assist patent owners with the prosecution and development of their patent portfolios, the protection of their patented inventions from unauthorized use, the generation of licensing revenue from users of their patented technologies and, if necessary, with the enforcement against unauthorized users of their patented technologies.' (Acacia Research Corporation, 2010).

Thereby, Acacia sees itself mainly as partner of research institutes, universities, small companies, and inventors who do not have the scale or the size to establish their own licensing program. To pursue this mission, *Acacia* contacts patent owners that have infringed or potentially infringed patents. If the patent owner is interested to assign the patents to *Acacia*, the patents are transferred to a LLC that is founded and held by *Acacia*. The patent aggregating company than takes action and follows a stick licensing approach by enforcing the patents in negotiations and litigation lawsuits. During the last years, the business model has started to shift from asserting only patents from single inventors and SMEs in single cases to long-term cooperations with MNEs to assert large patent portfolios. The advantage for *Acacia* is that MNEs have larger portfolios, with much deeper patent portfolios with a higher quality.

Organization. *Acacia* is a listed company with shares outstanding of USD 42.9 million.[24] Ninety-six percent of outstanding shares are free float; 145 institutional investors own 71.30% of the free float; and insiders own 3.88% of the free float.[25] The most important direct shareholders are CEO Paul Ryan and Director Robert Harris.[26] The top three institutional investors are *Eagle Asset Management* (4.83% of free float), *Vanguard Group* (4.52% of free float), and *Columbia Partners* (4.33% of free float).

Acacia holds more than 80 subsidiaries because for each licensing program, an LLC is formed. The company operates and finances its activities mainly from its own corporate treasury. *Acacia* formed a subsidiary together with institutional investors in August 2010 that serves as investment fund to acquire and license patents. Information on this fund is limited.

In addition to internal expertise, *Acacia* has a wide network of engineering experts and 30 law firms. Due to the wide range of targeted industries, external engineering experts are consulted for opinions and evaluations of technologies from their specific areas. As patent enforcement is the core pillar of revenue generation, external highly specialized law firms are employed for the enforcement of certain portfolios. Law firms also support *Acacia* in the due diligence process.

Process. *Acacia* has the resources to evaluate commercially valuable patents in any technology area or industry, but until now it has focused mainly on patents from high technology, for instance, communications; computers and peripherals; consumer electronics; digital media; ecommerce; energy and lighting; internet; medical devices; semiconductor; software; or wireless and mobile. *Acacia* focuses on the aggregation of US patents without additional knowledge or technology. At the beginning of its patent

---

[24] Status quo per September 6, 2011, data retrieved from Thomson Reuters. Closing stock price on September 5, 2011 of ACTG was USD 41.79.
[25] Status quo per September 6, 2011, data retrieved from Thomson Reuters.
[26] Reported on August 8, 2011, data retrieved from Morningstar, Inc.

aggregating activities, *Acacia* mainly aggregated patents from research institutes, universities, SMEs, and inventors. During the last years, it has increasingly aggregated patents from MNEs.

Acacia traditionally starts the structuring process. The patent aggregating company regularly monitors the patent landscape and identifies patents that are already used or are anticipated to be widely used by third parties. Based on the results of the monitoring activities, *Acacia* contacts the patent owners. During the first years of business activity, *Acacia* exclusively reached patent owners through patent attorneys and engineer screening. Becoming aware of patent infringements, licensing opportunities, and the existence of *Acacia*, patent owners increasingly approach the patent aggregating company and offer their patents.

Acacia follows a semi-standardized, two-stage evaluation and due diligence process that sorts out almost 97% of all patents or patent portfolios that were identified in the beginning. In the first stage of the due diligence process, *Acacia* conducts a preliminary analysis examining bibliographic data, general legal information, and the technology landscape. Only 10% of all patents or patent portfolios enter the second stage. In this stage, *Acacia* conducts a worldwide prior art search, performs reverse engineering, and identifies potential infringers. Additionally, all patents are evaluated from (1) a legal (e.g., is the patent well drafted, does it withhold the scrutiny of litigation); (2) an engineering (e.g., prior art of different types of technology, technology lifecycle, technological standards); and (3) a licensing and business perspective (e.g., profit margins, reasonable licensing rates, revenues of covered products). Based on experienced multidisciplinary teams, *Acacia* has the resources to conduct the evaluations and reviews internally. However, for specific technologies, external experts, such as law firms or engineers are employed. From the patents that have entered the second stage of due diligence, only 70% are viable for *Acacia's* exploitation strategy and are acquired by the company.

The contractual agreement *Acacia* enters with the original patent owners is internally called a 'partnering arrangement'. Through LLCs, which are formed for each licensing program, *Acacia* acquires the patent or the patent portfolio, with title changes at the US patent office, or it acquires the exclusive right to license a patent portfolio. In exchange, the original patent owner, called a 'partner', receives an upfront payment for the purchase of the patent portfolio or patent portfolio rights, a percentage of the net recoveries from the licensing and enforcement of the patent portfolio, or a combination of the two. Based on this arrangement, *Acacia* and the original patent owner equally share the economic value of the patent.

To exploit the patents, *Acacia* follows a stick licensing approach. Potential users of the patents are approached to meet and negotiate the case and enter litigation lawsuit only if negotiations cannot be closed or *Acacia* directly sues the infringing companies and starts a litigation lawsuit. The targeted result of both approaches is an agreement on a non-exclusive license, either to a specific portfolio or to all of *Acacia's* portfolios. In the past, *Acacia* has settled 95% of all litigation lawsuits out of court and rarely sells patents, and only if a major player wants to have an exclusive license. However, selling a patent must generate more revenue than licensing to the entire market.

Value for original patent owner. By selling patents to *Acacia*, the original patent owner can transfer enforcement risks to the patent aggregating company, as well as generate an immediate additional cash inflow. SMEs, universities, and single inventors often lack the resources and experience to enforce infringed patents. MNEs often do not want to enforce patents and prefer to stay out of certain litigation lawsuits. On the other side, producing companies can experience disadvantages from not enforcing patents. Selling the patents to *Acacia* resolves this problem. The patent owner receives an upfront payment and participates on the royalties if *Acacia* is able to enforce the patents.

5.4.3 Defensive patent aggregator's characteristics

Defensive patent aggregators acquire patents to provide the attached producing companies with an insurance against patent litigation lawsuits initiated from NPE. It generates revenues through membership fees and patent selling but not through patent enforcement. Therefore, patents are the only means to fulfill the requirements of the members. Defensive patent aggregators compete with patent enforcement companies for the same patents.

Defensive patent aggregators are the youngest category of patent aggregating companies and have emerged only recently. Therefore, the three companies in the sample cover 100% of the actual population. In this population, two major funding schemes are observable. Either defensive patent aggregators are established as an interest group of large producing companies and are financially backed by these founding members, as well as additional members, or they are privately founded and backed by large financial resources as venture capital firms. Depending on the funding scheme, defensive patent aggregators are not for profit or profit oriented. All business models are member based. To profit from the defensive patent aggregator, a company becomes a member and pays an annual fee. Depending on the defensive patent aggregators and the amount of fees, the annual fees are used to acquire patents or to cover the administrative costs. In the latter case, the acquired patents are paid from extra fees.

The senior management of defensive patent aggregators has experience in patent management and patent transactions gained in prior work for patent intermediaries or high-technology companies. Defensive patent aggregators often employ patent intermediaries to preserve anonymity and to obtain a realistic price in buying transactions.

Defensive patent aggregators focus on patents covering technologies in the high-technology industry, such as semiconductor, software, information and communication technology, wireless, and consumer electronics, but are not limited to these technologies. They aggregate patents already used by their attached producing companies that could become a threat if another company buys these patents, for instance, patent enforcement companies, patent acquisition companies, or patent trading funds. Defensive patent aggregators amass only the legal right of exclusion without any underlying knowledge. Until now, defensive patent aggregators have acquired mainly US patents.

Defensive patent aggregators acquire patents from single inventors, research institutions, SMEs, and MNEs, but also from other patent aggregating companies, such as patent acquisition companies, patent enforcement companies, or patent trading funds. Defensive patent aggregators take over the enforcement risks from producing companies. Compared with selling patents to corporate buyers, defensive patents funds can place bids higher than the willingness to pay for a single company. This is due to their membership structure and the joint fund of corporate buyers. At the same time, the original patent owner can enforce the patent without risking losing the lawsuit or investing the large amount of money affiliated with a patent litigation lawsuit and without getting in touch with patent enforcement companies or patent acquisition companies. Avoiding transactions with these categories of patent aggregating companies deletes the risk of reputation damage by selling it to a publicly unpopular company. By selling infringed patents, the producing company can generate additional short-term cash flows through the actual purchase price. Additionally, the producing company is able to save the costs of litigation.

The main objective of a defensive patent aggregator is to prevent patent litigation against its members. Therefore, patents that hold legal exposure to one or more of the members are acquired. The patent market is monitored constantly and available patents are bought. Depending on the business model, the member companies can influence the purchase decisions or not. The attached producing companies receive non-exclusive licenses from the aggregated patent portfolios. Interested companies can join the defensive patent aggregator if relevant patents are acquired. After a time span of around one to two years to give non-member companies the possibility to join and provide the members with a license, some companies sell the patents to other producing companies or to patent enforcement companies, patent acquisition companies, or patent trading funds. In the summary (Figure 35), the cash flows between the involved parties and the transfer of the patents is illustrated.

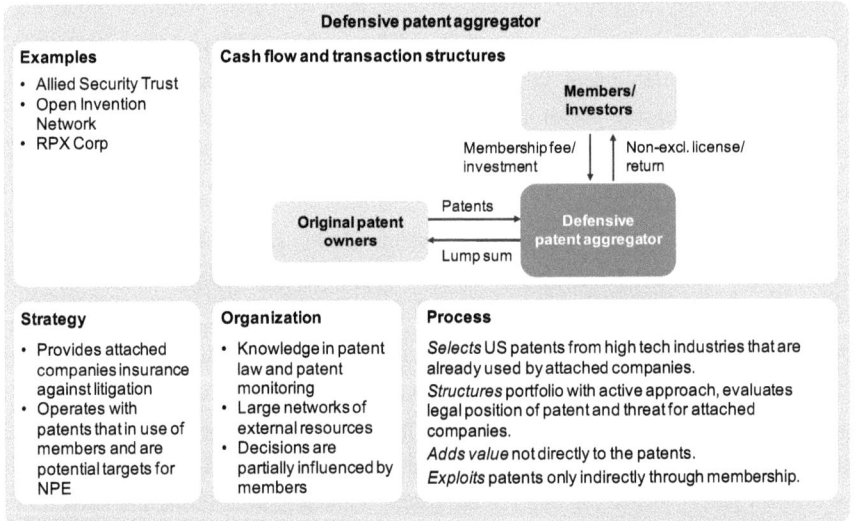

Figure 35: Summary of defensive patent aggregators

The unique characteristic of a defensive patent aggregator is that by aggregating patents and members, it is able to reduce the high costs of monitoring the patent market and the costs of acquisition of patents that hold large legal exposure for a single company.

5.4.4 Defensive patent aggregator's case study: Allied Security Trust

Setting. *Allied Security Trust* is an US patent defensive organization with headquarters in Lambertville, New Jersey. In January 2007, four companies from the high-technology industry founded *Allied Security Trust* as a company that should help the member companies to analyze and potentially purchase patents that may otherwise be used against the members in some type of aggressive action. Amongst the current members are *Ericsson, Hewlett-Packard, IBM, Intel, Motorola, Oracle,* and *Philips*. *Allied Security Trust* currently has 21 members, from Europe, North America, and Asia.[27] To become a member, a producing company must operate in the high-technology field, that includes software, information technology, communication technology, consumer electronics, internet, or medical devices, and generate operating revenues from production and services of USD 500 million per year. *Allied Security Trust* is open to new members. To join, new members have to pay a onetime fee of USD 150,000. *Allied Security Trust* is not a charitable organization but was designed without the objective to generate profits. It sees itself more as a trade association that serves the needs of its members. According to its charter and the trust agreements, the company is prohibited to assert patents.

Strategy. The mission of *Allied Security Trust* is to support members in dealing with aggressive actions performed by patent enforcement companies or patent acquisition companies. For this, *Allied Security Trust* identifies patents, purchases these patents on behalf of the member companies, therefore, taking member-threatening patents from the market. After purchasing and licensing the patents to the specific members, *Allied Security Trust* sells the patents to producing companies, as well as all kinds of patent aggregating companies.

Organization. *Allied Security Trust* is a member-owned trust. The annual membership fee for each member is USD 200,000. This fee covers all costs of the operating business of *Allied Security Trust*. Additionally, the members pay a certain amount for

---

[27] Status quo August 2011.

the patent acquisitions relevant for them. This amount is determined by each member separately depending on its own evaluation of the patents into question. *Allied Security Trust* works together with a worldwide network of patent brokers and patent intermediaries that offer patents to the company. Additionally, a network of experts supports *Allied Security Trust* upon request, providing third-party reports and opinions on certain patents or patent portfolios.

Process. Allied Security Trust members evaluate single granted patents, patent applications, and patent portfolios with up to thousands of patents from high-technology technologies that are offered at the market and could be turned into a litigation threat for the member companies. Often *Allied Security Trust* bids against patent aggregating companies as patent enforcement companies, patent trading funds, or patent acquisition companies, as well as operating companies. The acquired patents are mainly US patents and their foreign counterparts, and come from a worldwide network of over 300 brokers, operating companies, law firms, academic institutions, individual inventors, and other patent holders with patents for sale.

The organization monitors the market for patents constantly. The patent seller who offers patents to the trust mainly initiates the structuring process of *Allied Security Trust*. *Allied Security Trust* analyzes the offered patents regarding the relationship of the selling agent, family members, foreign counterparts, and fitting technology areas. All patents are evaluated taking the position of a potentially enforcing company. In the offered patent of interest, *Allied Security Trust* classifies the patents by two criteria: (1) potential products that may use the patent and that could be the basis for litigation lawsuits (e.g., routers, digital cameras, web-browsing technologies), and (2) the technology that the patent covers (e.g., antennas, imaging, liquid crystal). This information is delivered to the member companies that evaluate the patent and decide about the purchase. Based on the classification, *Allied Security Trust's* advice and their own investigation, the members decide if the patent meets their specific technical, product, and quality interest. In this case, each member company conducts an evaluation of the patent for itself and based on the evaluation, decides to participate in a bid and states the amount it is willing to pay for the patent. *Allied Security Trust* gathers all decisions and coordinates the process without disclosing the identity of the participating members. When the members are interested to make a bid, *Allied Security Trust* forms an LLC for each transaction. Based on the gathered amount, *Allied Security Trust* places an offer to the patent seller. If the offer is too low and the patent seller refuses to sell, *Allied Security Trust* contacts the interested members and, if the members are still interested, gathers additional funds and increases the bid. After acquiring the patents, *Allied Security Trust* grants a license to the members that funded the acquisition. On average, 20 to 30% of the members participate in a given acquisition. *Allied Security Trust* does not disclose the companies involved. The license granted is fully paid up, perpetual, irrevocable, worldwide, and non-exclusive. In addition, members that did not participate in the patent acquisition can obtain a license. In this case, they receive a 'Subsequent License Option' and pay the highest price paid by a member in the acquisition transaction. The license proceeds generated by new licensee are returned to the original bidders.

Figure 36 illustrates the process conducted in the structuring phase of *Allied Security Trust*.

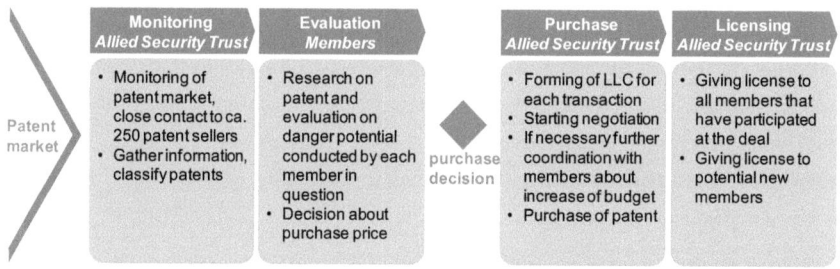

Figure 36: Structuring phase of Allied Security Trust

To prevent a free rider problem and secure the status as defensive organization *Allied Security Trust* divests the patents ca. one year after purchase. The patent portfolio is offered first to the original bidding members. If none of the original bidders is interested in purchasing the patent portfolio, the patents are offered to third parties. These third parties are producing companies and any other interested parties, whether they are patent aggregating companies or non-practicing entities. The proceeds received from selling the patents are transferred to the original bidding members, and they recoup some of their initial investments. To date, *Allied Security Trust* has returned 96% of the initial investment to the involved member companies.

Value for original patent owner. By selling patents to *Allied Security Trust*, the original patent owner generates an immediate additional cash inflow. In addition, the original patent owner can stay out of litigation lawsuits but at the same time react to infringements. Especially SMEs, universities, and single inventors often lack the resources and experience to enforce infringed patents. MNEs often do not want to enforce patents and stay out of certain litigation lawsuits. On the other side, producing companies can experience disadvantages from not enforcing patents. Selling the patents to *Allied Security Trust* resolves this problem. Even though *Allied Security Trust* does not assert the patents, the original patent owner receives rent from its innovation without initiating a litigation lawsuit. Additionally, *Allied Security Trust's* philosophy is to pay market prices. The original owner receives an appropriate upfront payment and does not have to cover renewal costs.

5.5 Archetype 4 – Patron

The archetype patron features nuisance competency and provides patent owners with monetary and non-monetary long-term rewards. As patent pooling companies and non-commercial patent aggregators evaluate and offer solutions for the legal cases of patents and reward patent owners with continuous payments, the transfer of litigation risks, and marketing tools, these two business models represent the archetype patron.

5.5.1 Patent pooling company's characteristics

A patent pooling company aggregates patents to solve licensing issues in technologies that are characterized by a large number of patents owned by several patent owners. A patent pooling company focuses on revenue generation for the patent owners but itself receives only cash flow for its administrating function. Patents serve as means to standardize technologies, as well as to solve the problem of patent thickets.

Patent pooling companies set up patent pools by or with the help of producing companies, because standardizing a technology or using a standardized technology is of major strategic interest for them. Patent pooling companies mainly administer the pooled legal rights and do not buy patents. Hence, large scale funding for patent acquisition activities is not necessary.

The senior management of patent pooling companies has broad experience in patent licensing and patent management.

Patent pooling companies focus on essential patents covering the basic technology of a standard. Until now, they are mainly active in the high-technology industry but the focus is slowly shifting to other industries, such as chemical, biotechnology, or manufacturing. Patent pooling companies aggregate only the legal right of exclusion without any underlying knowledge.

All original patent owners have the same objective for getting involved with a patent pooling company. Based on this, the original patent owners transfer certain rights concerning their patents to the patent pooling company. Original patent owners are single inventors, research institutions, SMEs, and MNEs. Patent pooling companies take over the enforcement risk from producing companies. Instead, enforcing patents that are intertwined with other patents and starting an 'endless' litigation game between all patent owners, a producing company can transfer the patents to a patent pooling company that consolidates all relevant patents. Assigning patents to a patent pool generates steady long-term royalties for the patent owners without having the costs to administer many licenses in-house. Additionally, the assigning patent owner can also receive a license for the patent pool, generate freedom to operate its own R&D, and has the opportunity to expand the market for its products because the developed technology is more widely used.

The patent pooling company acts as an administrator of an agreement between two or more patent owners. These patent owners have patents covering a standardized or to be standardized technology and plan to license them on a broad scale. The patent pooling company has two major functions. On the one hand, it aggregates additional patents to complete the patent portfolio for a standardized technology. On the other hand, it out-licenses the patent portfolio on a non-exclusive basis to producing companies of all sizes and collects the royalties. After deducting its administration fee, the patent pooling company passes the royalties to the original patent owners. In the summary (Figure 37), the cash flows and the transferred licensing rights between the involved parties are illustrated.

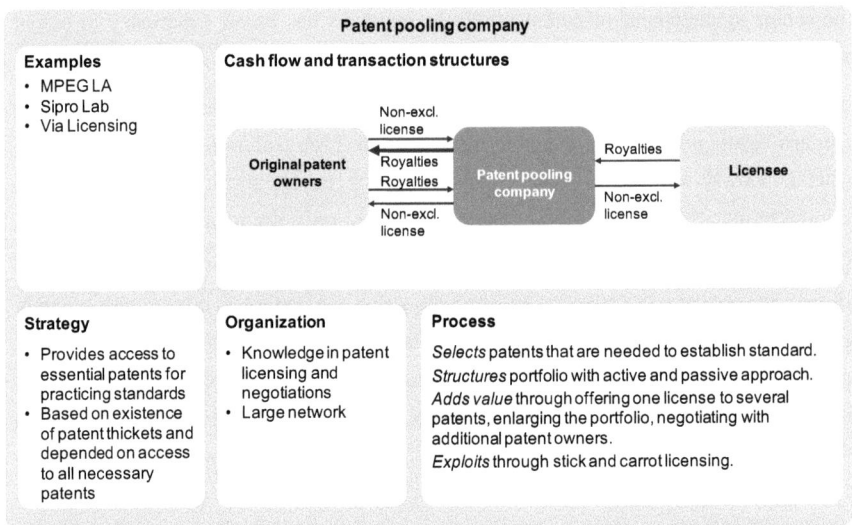

Figure 37: Summary of patent pooling companies

The unique characteristic of a patent pooling company is that by aggregating patents, it creates the opportunities to reduce transaction costs, to generate freedom to operate, and to expand markets for producing companies in one single transaction.

5.5.2 Patent pooling company's case study: MPEG LA

Setting. MPEG LA is an US patent aggregating company. MPEG LA was founded in 1997 as result of the Moving Picture Coding Expert Group's (MPEG) standardization attempts for coded representations of moving pictures, audio, and their combination. In 1992, the MPEG established the MPEG-1 standard, which is designed to produce reasonable quality images and sound at low bit rates. After further development, the MPEG-2 standard was established in 1994. This standard is designed to produce higher quality images at higher bit rates. The main problem was that many patents of several patent owners were essential to access the MPEG-2 technology. Due to the large number of necessary licensing negotiations, an adoption of the MPEG-2 standard seemed unlikely. Ken Rubenstein, co-head of the Patent Law Group of *Proskauer Rose* and a Partner in the New York office, was mandated to identify all relevant patent owners for the MPEG-2 standard. After identification and months of discussions and negotiations, the patent owners announced their licensing terms in March 1995. As a separate and independent entity that licenses the patents in and out and decides whether patents are a standard essential, *MPEG LA* was founded. *MPEG LA's* operations started in 1997. The MPEG-2 patent pool had initially 8 licensors and 25 patent families. Today the MPEG-2 patent pool has aggregated 710 patents in 57 countries from 27 patent owners. The patent pool is licensed to 1835 licensees. *MPEG LA* also administers six other patent pools: ATSC Broadcast standard – includes 196 patents from eight patent owners, 133 licensees; AVC/H.264 (MPEG-4 part 10) Video – includes 1733 patent from 27 patent owners, 1227 licensees; VC-1 Video – includes 519 patents from 18 patent owners, 236 licensees; MPEG-4 Visual (part 2) Video – includes 981 patents from 29 patent owners, 1086 licensees; MPEG-2 Systems (w/o MPEG-2 Video) – includes 204 patents from ten patent owners, 171 licensees; IEEE 1394 High Speed Data Network – includes 270 patents from 10 patent owners, 232 licensees. Four more patent pools are in formation.

Strategy. The mission of *MPEG LA* is to aggregate a portfolio that comprises a large number of patents covering a certain technology held by many different patent owners that interfere with each other and make it difficult for all parties to use this technology. *MPEG LA* includes all patents necessary to practice the particular technology. The company offers a non-exclusive license of this patent portfolio to all companies that are interested in practicing the technology and distributes the royalties between the patent owners. *MPEG LA* is a service provider to licensors and licensees and serves as administrator.

Organization. *MPEG LA* is a private company. The headquarters of *MPEG LA* are in Colorado. The company has offices in Washington D.C., London, Tokyo, and Shanghai. The internal competencies of *MPEG LA* focus on the relationship management with licensees, as well as licensors and the financial administration, communication, auditing, and reporting of royalties. Independent experts are employed for the evaluation and another opinion regarding the essentially of the patents to the specific technology. Additionally, external law firms handle the legal issues.

Process. *MPEG LA* aggregates patents covering a technology that faces the problems of patent thickets and that are interesting for a large group of companies. *MPEG LA* focuses on international patents from computer, consumer electronics, telecommunications, and related high-technology industries. The patent aggregating company amasses only the legal right without any technology or know-how. To pursue its strategy of providing a precise license for accessing the defined technology, *MPEG LA* aggregates mainly patents infringed by use of the defined technology. Patent owners involved with *MPEG LA* are also SMEs but mainly MNEs from the electronic high-technology industry. Additionally, research institutions, such as *Fraunhofer Gesellschaft*, have assigned rights to license to *MPEG LA*.

MPEG LA follows an active approach to structure the patent portfolio. Based on their webpage, the offices in Europe and Asia and their network of service providers *MPEG LA* releases a 'call for essential patents' when a patent pool is formed. Patent owners react to this call and offer their patents to *MPEG LA*. To evaluate the essentially of a patent for the defined technology, independent patent experts are employed. Kenneth Rubenstein of *Proskauer Rose* heads the independent patent evaluation and is *MPEG LA's* US patent counsel. Other members of the team include Gottfried Schull, Thomas Rox, and Ralph Schippan of *Cohausz & Florack* in Düsseldorf for the evaluation of European patents; Hideo Ozaki of *Ohba and Ozaki* in Tokyo for the evaluation of Japanese patents; and *Moon & Moon* in Seoul for the evaluation of Korean patents. If the offered patents are suitable for the patent pool, licensors are required to include all of their essential patents.

MPEG LA gives companies the opportunity to practice a certain standard. Therefore, *MPEG LA* offers interested companies a license to a patent pool that consists of all essential patents to use the standard. The various sublicenses granted by the license are worldwide, non-exclusive, and non-transferable. The licensee pays royalties for each produced unit that uses the patents from the patent pool. The license reflects a balance of royalty, revenues, and administrative fees. For the MPEG-2 patent pool, *MPEG LA* has adjusted the royalty rate four times. Due to changing business conditions, the royalty rate was reduced every time. *MPEG LA* treats the data of licensees confidentially and does not disclose the names to patent owners, other licensees, or third persons. *MPEG LA* refers always to its administerial function and does not enforce the patents aggregated in the patent pool. If companies use patents of the MPEG-2 standard and do not agree to take a license, *MPEG LA* informs the patent owner about the situation. The patent owner decides whether to take action. Even licensing thousands of patents to several thousand companies, *MPEG LA* has notified patent owners about infringements less than 40 times.

Value for original patent owner. By assigning patents to *MPEG LA*, the original patent owner receives cash flows without the internal costs of licensing negotiations, licensing programs, and licensing audits with several licensees. Additionally, *MPEG LA* regularly evaluates patents that could be added to the patent pool. That increases the adoption of a technology and could increase revenues from royalties and sales of own products.

5.5.3 Non-commercial patent aggregator's characteristics

A non-commercial patent aggregator amasses patents and technologies to neutralize licensing issues in the areas of social or humanitarian importance and makes patents available for a broad range of users. A non-commercial patent aggregator does not focus on revenue generation but on fostering innovation and social welfare. Patents and technologies serve as means for its fostering activities.

Often patents are donated to non-commercial patent aggregators. If a non-commercial patent aggregator intends to buy patents, public authorities, non-profit organizations, or companies with major interests in the non-commercial patent aggregator fund the acquisition activities.

The senior management of non-commercial patent aggregators has a technical or general management background, often applied in research institutions, development aid agencies, or other public bodies. External resources as licensing agents, patent attorneys, patent lawyers, patent intermediaries, or engineers are employed to identify and select the patents and to make them available for users.

Non-commercial patent aggregators are always set up for a special purpose and therefore, focus only on patents serving this purpose. Single interests do not drive the special purposes but they are intended to serve the public, underprivileged groups, or ecological development. Non-commercial patent aggregators aggregate the legal right of exclusion and in certain cases, the underlying knowledge. Depending on the purpose, the geographical application of the patents varies, and, for instance, includes only patent documents granted in developing countries.

As the targeted patents are carefully selected and specifically applied, the original patent owners are diverse. Depending on the purpose of the non-commercial patent aggregator and the area the targeted technology is located in, they can range from research institutions and universities to MNEs. Non-commercial patent aggregators take over the enforcement risk from producing companies. Instead of enforcing patents that cover areas of high public visibility and interest, a producing company can transfer the patents to a non-commercial patent aggregator. On the one hand, it can prevent suing social welfare increasing projects and risking damage to its reputation. On the other hand, the companies circumvent the problem of not enforcing its patents. Often producing companies donate patents to non-commercial patent aggregators and can claim a tax deduction.[28] Additionally, the donating company can use the donation as a marketing tool and save R&D costs. Applying the donated patents and innovations on this basis, the donating company receives access to the new inventors. If patents are essential for the functioning of the non-commercial patent aggregators, patents are also acquired or exclusively in-licensed.

A non-commercial patent aggregator follows its mission and does not focus on profit generation. Based on this mission, patents that serve the targeted purpose are identified, and the non-commercial patent aggregator or an external service provider contacts patent owners. The only relevant determinant in the evaluation process is the fit with the non-commercial patent aggregator's mission. Having aggregated the patents, the non-commercial patent aggregator administrates the patent portfolios and enlarges them. In general, non-commercial patent aggregators administrate the non-commercial licensing program, but they do not enforce patents or exploit them in a commercial way. In the summary (Figure 38), the relationships between the original patent owners, the non-commercial patent aggregator, and the patent users are illustrated. As the figure shows, in general, only cash flows as financial support are transferred.

---

[28] In the US, donating patents to a non-profit organization can reduce taxes. The donating company is allowed to claim either the market value or the R&D costs that were necessary to develop the patent as a tax deduction. Additionally, a percentage of revenues created with products from the donated patent possibly can be deducted for up to 12 more years. For further information on charitable donations of patents, see IRS Publication 526.

Archetype 4 – Patron 147

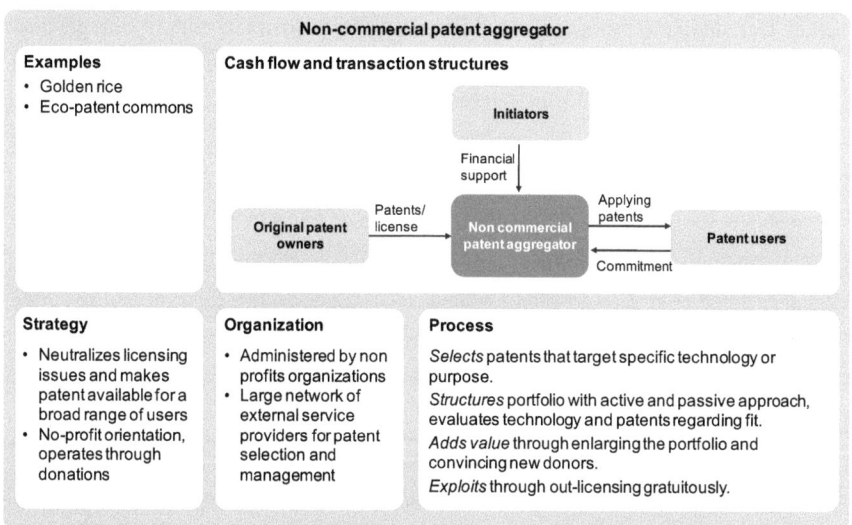

Figure 38: Summary of non-commercial patent aggregators

The unique characteristic of a non-commercial patent aggregator is that by aggregating patents, it is able to follow a special purpose without profit orientation.

5.5.4 Non-commercial patent aggregator's case study: Golden Rice PDP

Setting. The *Golden Rice product development partnership (Golden Rice PDP)* is a result of an initiative of the *Rockefeller Foundation* accomplished by Professor Ingo Potrykus of the *Institute of Plant Sciences* at the *Swiss Federal Institute of Technology* and Professor Peter Beyer of the *University of Freiburg*. The project started in 1992 and scientific results were first published in the journal 'Science' in 2000. Due to malnutrition, large parts of the population in less developed countries in Asia and Africa suffer from chronic vitamin A deficiency. Vitamin A deficiency is responsible for 1 to 2 million deaths, 500,000 cases of irreversible blindness, and millions of cases of xerophthalmia annually, which leads to night blindness (Humphrey, West Jr., & Sommer, 1992). The developed Golden Rice is genetically modified rice that produces high level of beta-carotene, a precursor of vitamin A. The newly developed rice could help to reduce vitamin A deficiency-related diseases.

The key technology to produce Golden Rice is patented by the inventors who planned to provide the technology free of charge and restrictions to farmers and research institutes in developing countries. Problems arose when Golden Rice planned to be commercialized because in the production of Golden Rice, 70 patents and patent applications from 32 patentees are involved. To solve this problem and make Golden Rice available for humanitarian use, the *Golden Rice PDP* was established.

Strategy. The mission of *Golden Rice PDP* is to transfer and introduce the newly developed breed of rice to developing countries. To pursue this mission, *Golden Rice PDP* gives access to the technology and the pertaining patents for defined humanitarian research and the use of Golden Rice for subsistence farmers in developing countries free of charge.

Organization. The general decision on licenses and strategic use of the Golden Rice technology is made by the *Golden Rice Humanitarian Board.* This board consists of representatives of *Syngenta, Rockefeller Foundation, the World Bank*, public and private research institutes, as well as the inventors. It aims to facilitate further Golden Rice research and to introduce Golden Rice to developing countries. The patents and licensing agreements are administered within the global Swiss agribusiness company *Syngenta.* The *Golden Rice* Project Manager is Dr. Jorge Mayer of *Campus Technologies Freiburg, University of Freiburg*, Germany. The Golden Rice Network initially deploys Golden Rice. This network consists of research institutions and universities from the Philippines, India, Vietnam, Bangladesh, China, Indonesia, and Germany. Dr. Gerard Barry of the *International Rice Research Institute* from the Philippines coordinates it. The institutions of the *Golden Rice Network* are responsible for introgressing the Golden Rice trait into local varieties.

Process. The *Golden Rice PDP* focuses on patents that cover the technology to produce the newly developed rice breed Golden Rice. Because the core technology was already developed, only the patents are of interest without any underlying knowledge or technology. A freedom to operate analysis was conducted when the technology was ready to be further developed for humanitarian purposes and commercialization. The freedom to operate analysis, conducted by the *International Service for the Acquisition of Agri-biotech Applications*, showed that 70 patents and patent applications from 32 companies and universities were applicable to the newly developed breed of rice. The result of further analysis was that 11 patents had a high potential to serve as a barrier to the deployment of Golden Rice in countries with the highest levels of vitamin A deficiency. Therefore, these patents that could prevent deployment were automatically the targeted patents. The selection of patents, as well as of patent owners only focused on the objective to generate freedom to operate for the Golden Rice technology.

In the structuring phase in 2000, the *Golden Rice PDP* between the inventors and *Syngenta*[29] was established. Facing the challenge that patents could restrict Golden Rice and the complex negotiation for licensing agreements, the inventors approached a number of patent owners to find a private partner. On May 16, 2000, *Syngenta* announced the collaboration to make rice containing vitamin A available free of charge for humanitarian use. *Syngenta* was involved in a research project funded by the *European Commission* of which Golden Rice was a small part. The company holds patents that cover technologies necessary to produce Golden Rice from this project. To aggregate the patent portfolio, the inventors assigned their exclusive rights to *Greenovation,* a spin-off of the *University of Freiburg* that out-licenses university biotechnology research projects. *Greenovation* assigned the exclusive rights to *Syngenta.* The aggregation of the patent portfolio was completed when *Syngenta* gave access to all patents of *Syngenta* and *Syngenta Seeds* and negotiated access to the related patents of *Bayer AG, Monsanto, Novartis, Orynova,* and *Zeneca Mogen.* All companies provide access to their technology free of charge for humanitarian research and the use of Golden Rice in developing countries. Based on positive publicity in TIME magazine, *Monsanto* offered a free license to the inventors. The increasing public pressure led to the agreements with the other companies. Figure 39 illustrates the patent aggregating process and the involved parties.

The research project developed the Golden Rice technology and provided a proof of concept but did not develop marketable products. In the value-added phase, *Syngenta* developed the proof of concept results into deliverable products.

---

[29] The partnership was established with *AstraZeneca*. On November 13, 2000, *AstraZeneca* merged with the seed and agrochemical division of *Novartis* to form *Syngenta*.

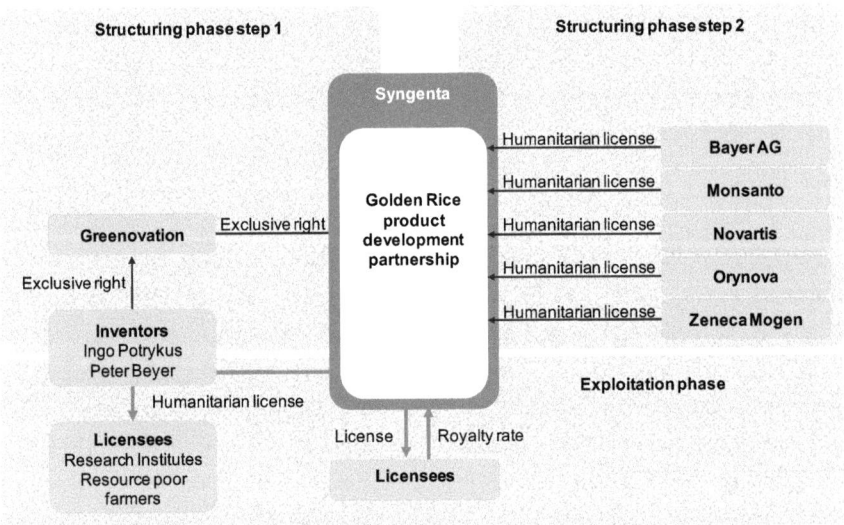

Figure 39: Patent aggregating process of Golden Rice PDP

The patents of the *Golden Rice PDP* are exploited in two ways. *Syngenta* has the full commercial rights to the invention worldwide and can therefore commercialize Golden Rice in the developed world. In the beginning of the partnership, *Syngenta* estimated a market for health conscious customers in industrial nations and planned to introduce Golden Rice as nutritionally enhanced food. In 2005, *Syngenta* decided not to go commercial with Golden Rice in developed countries. Additionally, *Syngenta* issued a humanitarian license to the inventors who have the right to sublicense the *Golden Rice* technology to national and international research institutes and resource poor farmers in developing countries free of charge. *Syngenta* also has the right to license for humanitarian use.[30] The *Golden Rice Humanitarian Board* is the body that gives advice on all issues and licensing agreements regarding the humanitarian use.

Value for original patent owner. By giving access to the technology for humanitarian use, the original patent owner can prevent the enforcement of these patents. Enforcing patents used in humanitarian projects can damage the reputation of the company.

---

[30] The research project developed the Golden Rice technology and provided a proof of concept but did not develop marketable products. In the value-added phase, Syngenta developed the proof of concept results into deliverable products.

Getting involved in non-commercial patent projects can prevent that and additionally generate long-term benefits through reputation enhancement. *Golden Rice* is a highly visible project. Assigning patents that help to improve conditions for people in developing countries can be used for positive public relations, as well as counter-argument in cases of critics on patenting, especially in the life science industry. In particular, *Syngenta* has profited from the *Golden Rice PDP*. By assigning the patents of the inventors to *Syngenta* and the right to the worldwide commercial use, the company received a new almost marketable technology, financed by public funds, with little R&D effort. Research institutes that develop the technology further have to transfer the commercial rights of improvements to the technology to *Syngenta*.

5.6 Summary and evaluation of potentials

The analysis of the 27 case companies finds that patent aggregating companies differ substantially regarding their strategies and motives to buy patents. That allows eight different business models to be distinguished. The analysis also finds that patent aggregating companies can be clearly distinguished regarding their competencies and the rewards they offer to the original patent owners. Based on these results, four archetypes are identified:

(1) The Merchant – this archetype features business competency and provides patent owners with monetary short-term rewards, such as lump sum payments.
(2) The Gardener – this archetype features business competency and provides patent owners with monetary and non-monetary long-term rewards, such as continuous payments to improve the financial situation, organizational learning opportunities, the transfer from commercialization risks, and insurance against losses of future cash flows.
(3) The Collector – this archetype features nuisance competency and provides patent owners with monetary short-term rewards, such as lump sum payments.
(4) The Patron – This archetype features nuisance competency and provides patent owners with monetary and non-monetary long-term rewards, such as continuous payments, marketing tools, the transfer of litigation risks, and marketing tools.

Two business models represent each archetype. The business models differ regarding the breath of transaction. For each archetype, one business model focuses on the aggregation of the sole legal right of exclusion. The other business model amasses patents but also technologies, and knowledge.

The four archetypes of patent aggregating companies are able to realize the aforementioned external and internal potentials to different degrees. Based on the empirical findings and reflecting them on the typology and the potentials, Table 3 evaluates the different potentials by business model. The business models have a different breath of transaction. Therefore, they differ slightly in the potentials they offer. To describe the potentials accurately, the business models rather than the archetypes are evaluated.

In line with the business competency they offer, the archetypes merchant and gardener can realize particular external potentials. Their business models are based on market knowledge and technology understanding. Therefore, the external potential of market interaction can be realized, in contrast to the archetypes that only work with nuisance competency.

In addition, the archetypes merchant and gardener are able to take over R&D risks. Both archetypes amass patents that have a business case. Even though directly realized in commercialized products, the original patent owner is able to recoup investments from R&D. In certain cases, not only the past risks of R&D but also future R&D risks are transferred. The archetypes collector and patron take over enforcement risks in a certain way and offer the original patent owner an alternative way to patent litigation.

		External potentials			Internal potentials		
	Market interaction	Market fostering	Resource enhancement	Risk reduction	Cost effectiveness	Decision making
Merchant — Patent trading fund	Market penetration (e.g., Alpha Patentfond), Opportunity Identification (e.g., Alpha Patentfond)	Liquidity and market clearing (e.g., Alpha Patentfonds, Patent Invest)	Access to resources (e.g., Alpha Patentfonds, Patent Invest)	R&D risks' hedging (e.g., Alpha Patentfonds, Patent Invest)	Organizational learning (e.g., Alpha Patentfond), Transaction costs (e.g., Alpha Patentfond)	Innovation strategy (e.g., Techquity)
Merchant — Patent acquisition company	Opportunity identification (e.g., Intellectual Ventures)	Liquidity and market clearing (e.g., Intellectual Ventures)	Access to resources (e.g., Intellectual Ventures, Tequity)	R&D risks' hedging (e.g., Intellectual Ventures, Techquity)		
Gardener — Royalty monetization company	Market penetration (e.g., AlseT)		Access to resources (e.g., Capital Royalty, Paul Capital Healthcare)	R&D risks' hedging (e.g., AlseT, Royalty Pharma)		
Gardener — Patent incubating fund	Opportunity identification (e.g., IP Holdings)	Liquidity and market clearing (e.g., Patent Select)	Access to resources (e.g., IgniteIP, Patent Select), Network (e.g., Patent Select)	R&D risks' hedging (e.g., IgniteIP, Patent Select)	Organizational learning (e.g., Patent Select), Transaction costs (e.g., IgniteIP)	Company strategy (e.g., Patent Select), Innovation strategy (e.g., Patent Select)
Collector — Patent enforcement company		Liquidity and market clearing (e.g., Acacia)	Access to resources (e.g., IP Navigation, Papst)	Enforcement risks' hedging (e.g., Alliacense, Papst)	Transaction costs (e.g., Acacia)	
Collector — Defensive patent aggregator		Liquidity and market clearing (e.g., AST, RPX)	Access to resources (e.g., AST, RPX), Networks (e.g., OIN)	Enforcement risks' hedging (e.g., AST, RPX)	Transaction costs (e.g., RPX)	
Patron — Patent pooling company			Access to resources (e.g., Sipro Lab, MPEG LA, Networks (e.g., Via Licensing)	Enforcement risks' hedging (e.g., Sipro Lab)	Transaction costs (e.g., MPEG LA)	Company strategy (e.g., Eco-Patent Commons)
Patron — Non-commercial patent aggregator		Network (e.g., Eco-Patent Commons)		Enforcement risks' hedging (e.g., Golden Rice)		Company strategy (e.g., Eco-Patent Commons)

Table 3: Evaluation of potentials by business model

6 Leveraging patent portfolios by utilizing patent aggregating companies

Patent aggregating companies show eight characteristic business models and can be classified in four general archetypes, which are distinct in the rewards they pay the original patent owner and in the competencies they offer.

In the following chapter, a management framework is derived. This framework provides guidelines for managers of producing companies for which patents and circumstances for which patent aggregators can be utilized. As the patent aggregation business and the market for patents and technologies are fast changing environments, the management framework reflects the status quo and the recent strategies and activities of patent aggregating companies. To understand the dynamics and to assess the future direction of patent aggregating companies' business models, in the second part of the chapter, the development of patent aggregating companies is analyzed. In the third part, the driving factor behind the development is identified.

6.1 Managing the utilization of patent aggregating companies

Producing companies have several options to generate value from their patent portfolio (for an overview of value generating options, see Figure 5). Patent aggregating companies cannot be utilized for all of them. The following part discusses for which value generating options patent aggregating companies can be utilized and where the limitations of patent aggregating companies' utilizations are. In the last part, a management framework is presented that aligns the typology of patent aggregating companies, the value generating options they support, and the limitations producing companies face, and this serves as a guideline for patent managers to make the final decisions on the utilization of patent aggregating companies. As the patent aggregating industry is a fast changing industry and new business models emerge as established ones may vanish, the management framework reflects the current situation.

6.1.1 Value generating options and patent aggregating companies

As patent aggregating companies amass patents and usually acquire ownership rights or exclusive rights to exploit the patents, these companies can be utilized in external patent exploitation projects, more specifically for external patent exploitation projects that focus on donating, selling, or out-licensing patents. Figure 40 gives an overview of the different value generating options that can be conducted with patent aggregating companies and the business model of patent aggregating companies that can be utilized for the specific value generating option.

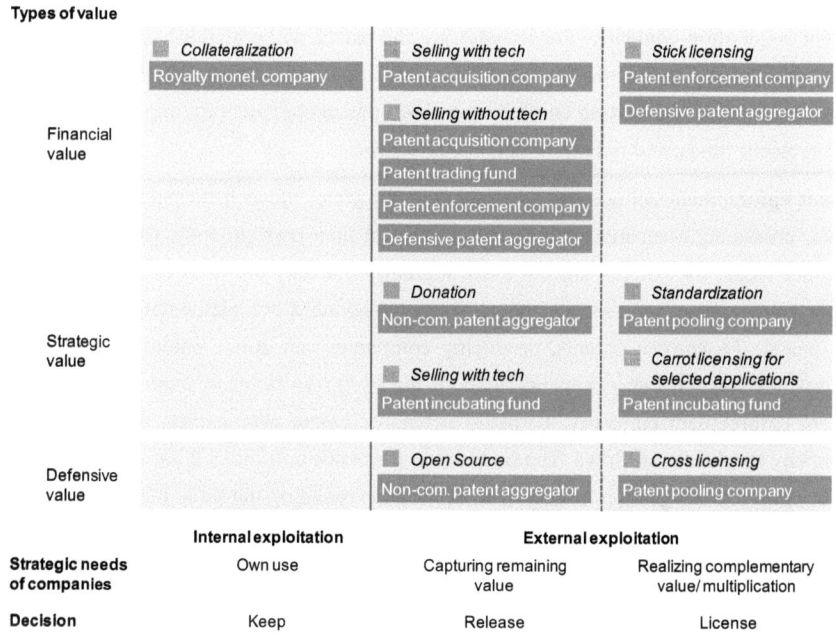

Figure 40: Value generating options that can include patent aggregating companies

For the objective of generating financial value from patents, producing companies can sell patents with or without technology or enforce infringed patents and generate licensing revenues from stick licensing. Additionally, patents that already generate revenues from internal use or existing carrot licensing agreements can be used for collateralization.

Patent owners can utilize royalty monetization companies for generating immediate cash flows from patents that generate long-term cash flows. The patents are used to protect a product internally (or they are already out-licensed). The resulting cash flows (true royalties or product cash flows) can be sold to the royalty monetization company and the patent, even used internally, can generate financial value.

Several business models of patent aggregating companies can be utilized for selling patents. Producing companies divest patents mainly for financial reasons. In transactions of patents and technologies, a producing company can sell the patents to a patent acquisition company. Transactions of the sole legal right can also utilize the patent acquisition company. For transferring the patent without additional technology and knowledge, the producing company can approach several different business models of patent aggregating companies and sell patents to patent trading funds, patent enforcement funds, and defensive patent aggregators.

Patent enforcement actions can also generate financial value from patents. In some cases, producing companies use patents to protect their own products from imitation, in other cases, the patents are not used internally but they are used to assert patents against infringers. Both ways, patents always generate a complementary value from the patent. To enforce patents, producing companies can utilize patent enforcement companies and sell the ownership rights or the exclusive rights of exploitation to the patent enforcement company. Utilizing defensive patent aggregators, the producing company receives cash flows. The transaction generates non-direct additional financial value from enforcement, but the generated value is based on the same mechanism, and the producing company is able to generate complementary value.

Besides the pure revenue generating aspect of utilizing patent aggregating companies, producing companies can use several business models of the patent aggregation companies to generate strategic value. An option to generate strategic value is the donation of patents. To create strategic value in the form of potential innovation inflows and from reputation and marketing effects, a producing company can utilize non-commercial patent aggregators. By donating patents to a non-commercial patent aggregator, the producing company releases a patent not internally used and captures remaining value from this patent.

A patent, in addition with technology and knowledge, can also generate strategic value if a producing company sells the patent to a patent incubating fund. In addition to the cash inflows from the sale, the producing company can commercialize a product in the long-term and therefore, secure or expand its market position. Learning effects and resource enhancements can back this development. Value from the multiplication of the technology can be realized if the producing company utilizes the patent incubating fund for a carrot licensing of the advanced technology.

A producing company with the strategic objective to standardize a technology, to secure its market position, and to gain new market shares can utilize patent pooling companies. In addition to the strategic value created from patents transferred to patent pooling companies, these transactions generate a defensive value. Patents in a patent pool serve as basis for cross-licensing agreements between players in a standard and prevent patent infringements in the standardized technology.

The option of releasing patents for open source projects generates defensive value. A producing company can utilize non-commercial patent aggregators for open source transactions and hence, generate access to other technologies and innovation without paying for it and, very importantly, without infringing patents.

6.1.2 Constraints in utilizing patent aggregating companies

Despite the potentials patent aggregating companies offer producing companies for leveraging their patent portfolios, several constraints apply for the utilization of patent aggregating companies. Hence, value-generating options are not always straightforward to realize. Based on a comprehensive analysis of literature and empirical data, four constraints can be derived. The four general constraints that affect the utilization of patent aggregating companies and the choice of value generating options are:

- *Value of patents*: Even though patent aggregating companies acquire vast quantities of patents, they are not interested in worthless patents.
- *Timing*: Most patent aggregating companies have specialized business models. They buy only patents at certain times, and they buy only patents covering technologies in a certain stage of the product lifecycle.
- *Industry*: Patent aggregating companies are mainly interested in markets and industries with a high relevance of patents and large revenue potentials.
- *Feed the troll*: In the last decade, a systemized and financial powerful patent enforcement industry has evolved. Selling patents to patent aggregating companies that focus on patent enforcement may fuel this system.

Figure 41 gives an overview of the four constraints that hinder the straightforward utilization of patent aggregating companies.

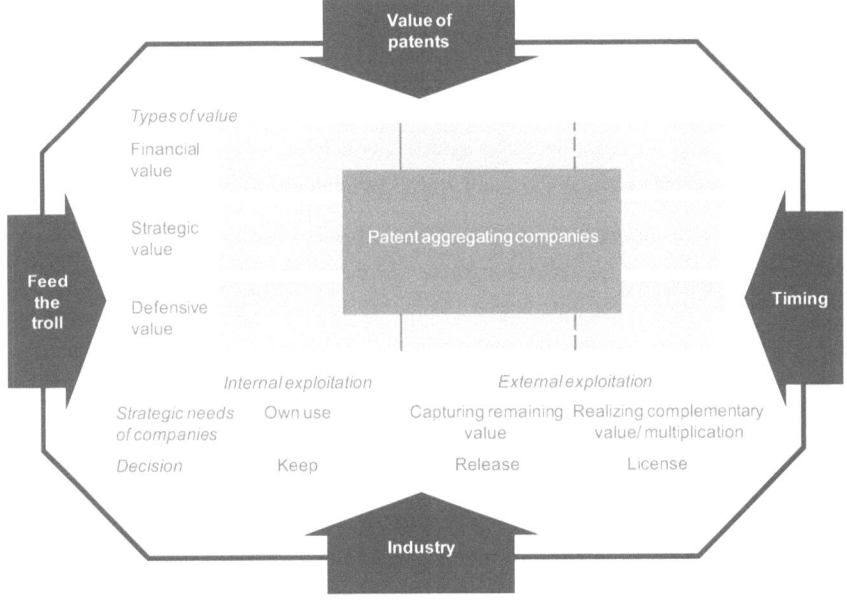

Figure 41: Constraints that affect the utilization of patent aggregating companies

Value of patents

The PatVal-EU project analyzed the value and use of more than 9,000 European patents. An important result of the study is that about 17.4% of all patents in a patent portfolio are so-called 'sleeping patents'. Sleeping patents are defined as patents not employed in internal use (protecting production processes or incorporated in a product), licensing, cross licensing, licensing and internal use, or blocking competitors (Giuri et al., 2007). Even unused, these patents may still have option value to the holder.

An accurate valuation of patents is still difficult to conduct but several studies confirm that the general distribution of patent values is highly skewed (e.g., Gambardella et al., 2008; Griliches, 1990; Harhoff, Scherer, & Vopel, 2003a, Harhoff, Scherer, & Vopel,

2003b; Scherer, 1965). That means that only a small proportion of patents has a high value. The majority of patents in a patent portfolio are on the low-value side and of little value. Schankerman and Pakes (1986) find that only 5% of all patents generate circa half the value of a patent portfolio. If the original patent owner tries to exploit these patents economically, they produce a zero or even negative stream of profits (Gambardella et al., 2007).

In general, producing companies are mainly willing to sell patents that are not in use, more specifically, they are willing to sell above-mentioned sleeping patents. If these sleeping patents result from research projects stranded due to strategy changes or from companies that were acquired but do not fit the new company's strategy, these patents may have a large economic value for third parties. However, only a few patents are on the valuable tail of the value distribution, these sleeping patents have little economic value for the patent owner, and often they have only little economic value for other companies. Instead of abandoning the sleeping patents, firms try to sell them and often see patent aggregating companies as buyers that acquire everything they are offered. However, patent aggregating companies do not serve as a garbage dump. They buy only patents with economical value. Even though patent aggregating companies acquire patents in large quantities from a broad range of industries and technologies, they have specific business models and focus on patents that can generate value in different forms. If the original patent owner does not see a market, now or in the future, for the technology, or the technology is already obsolete, it might be better to abandon the patent instead of investing internal costs to become engaged with a patent aggregating company. Some patent aggregating companies focus on buying embryonic or already commercialized technologies but they do not buy patents without, recent or future markets.

Timing

Original patent owners also face constraints regarding the timing. The first timing refers to the acquisition phase of the patent aggregating company. The second timing refers to the timing of selling patents in the right stage of the product life cycle.

Several patent aggregating companies do not amass patents all the time. Backed by financial investors, they first acquire money and then amass the patents. That means they have only a certain window of time for patent acquisition.

Products and technologies pass through different stages, ranging from their invention to their withdrawal from sales. This process is called the product lifecycle (Levitt, 1965). A generic product lifecycle can be divided into the two main phases: product development and product commercialization, and their six different stages. During the product development phase, the product does not generate revenues but the inventing company has to invest in R&D. Once the product is introduced to the market, it generates revenues. The amount of revenues and especially the annual growth rate of revenues depend on the age of the product. At a certain point in time, most products have reached their revenue peak, and sales start to decline.

Along with the product, lifecycle moves the technology lifecycle and the patents that cover the technology. Discovered in basic research or as idea in the development, the technology evolves from an embryonic technology to a legacy technology. In general, patents cover the developed technologies and change during the lifecycle regarding their claims, scope, and applications. Basic patents protect the results from basic research. As the technology advances, the number of patents and the number of patent applicants increase, different applications and substituting technologies are covered, and blocking patents are filed.

Producing companies make the decision to sell patents and to become engaged with patent aggregating companies in all stages of the product lifecycle. However, the success of utilizing patent aggregating companies strongly depends on the match between the patent aggregating company and the stage of the lifecycle the patent is located in. Patent aggregating companies specialize in amassing patents covering technologies from one certain stage of the product lifecycle. Patents that are filed to cover basic research are only interesting for patent incubating funds. These patent aggregating companies focus on conquering the stage of development and scaling up. Patent aggregating companies that enforce patents or trade patents have business models that are not able to exploit patents from embryonic technologies. Patents that cover products or technologies in the product commercialization phase and therefore, are already quite mature and easier to evaluate are interesting for a larger number of patent aggregating companies. Nevertheless, the actual age of patents is an important criterion for all patent aggregating companies. Patents close to their expiry date are not interesting for any patent aggregating company regardless of the product lifecycle stage.

Industry

The value of patents, the patenting behavior of firms, and the strategic use of patents differ across industries and technology classes (e.g., Ernst, 2001; Gassmann & Bader, 2011; Giummo, 2010; Giuri et al., 2007; Griliches, 1990; Lanjouw & Schankerman, 2001; Levin, 1986).

In the chemical and pharmaceutical industry patents are an important and effective means to protect innovations (e.g., Cohen et al., 2000; Levin et al., 1987; Mansfield, 1986). Patents are effective when the development of new products is expensive, but relatively cheap to imitate (Arundel & Patel, 2003). Studies show that in the chemical and pharmaceutical industry, about 80% of patentable inventions is patented (Mansfield, 1986). Patents are generally used to secure the market power of chemical and pharmaceutical products (Gassmann & Bader, 2011). In particular, blockbuster products in the pharmaceutical industry are highly dependent on patents, since generics produced by other firms constitute a high threat to the revenue created by the respective blockbuster product. The value of patents within the chemical and pharmaceutical industries, on average, is of higher value than patents from other industries (Giuri et al., 2007). Chemical and pharmaceutical industry differ substantially regarding litigation rates, since there is a case filed for 20% of the pharmaceutical patents, whereas the litigation rate for chemicals is very low. In general, pharmaceuticals are the most litigated technology group of all assessed groups (Lanjouw & Schankerman, 2001).

Seven of the top ten patent applicants at the EPO are companies from the technology field of electrical engineering. The companies *Philips* and *Siemens* are ranked number one and two with 2,556 and 1,708 patent applications, respectively in the year 2009 (EPO, 2009). A high interdependence between firms resulting from patents characterizes the electrical engineering industry. Companies are not able to market new products autonomously without being contingent on third-party patents (Blind et al., 2009). Especially in the telecommunication industry, the interdependence between firms is particularly distinctive (Gassmann & Bader, 2011; Leiponen & Byma, 2009). This results in cross-licensing actions, which are prevailing and inevitable for companies in this industry. As found by various researchers, the share of cross-licensing in electrical engineering is above average (e.g., Anand & Khanna, 2000; Giuri et al., 2007; Grindley & Teece, 1997; Hall & Ziedonis, 2001). Consequently, companies often only file for patents to block competitors (von Graevenitz, Wagner, &

Harhoff, 2008), or to strengthen their positions in cross-licensing negotiations (Cohen et al., 2000). Therefore, in the electrical engineering industry, patents are not an effective means to protect innovations. Companies rate other means to secure profits from R&D (e.g., a head start, establishment of effective production sales and service facilities, and rapid movement down the learning curve) as much more effective than patents (Mazzoleni & Nelson, 1998).

Based on the relevance of patents in the different industries and the patenting behavior of firms in these industries, producing companies are limited regarding their choice of patent aggregating companies. Often, business models of patent aggregating companies work only in specific industries. To enforce acquired patents, the patent has to be in use; infringement has to be easy to proof; infringers have to be easy to detect; there must be a sufficient number of infringers or one infringer with a very large market share and large revenues resulting from the infringed patents; and the outcomes of litigation lawsuits have to be difficult to predict. All these factors are mainly prevalent in the high-technology industry. In contrast, business models of patent aggregating companies that transfer technologies, or license patents for different applications have to operate in an environment where only few patents cover a product. The buyer or licensee of the patent portfolio can only make limited use of the patents if they are blocked by a web other legal rights. A general condition is that the market for the protected product offers potential for large revenues. Patent aggregating companies rarely buy patents covering technologies commercialized in niche markets or from niche industries.

Feed the troll

Peter Detkin first coined the term 'patent troll' in 2001. At this time, Peter Detkin (now Vice Chairman of *Intellectual Ventures*, a patent acquisition company) was Vice President and Assistant General Counsel of *Intel*. Using the term patent troll, Detkin described "a patent troll is somebody who tries to make a lot of money off a patent that they are not practicing and have no intention of practicing and in most cases never practiced" (as cited in Sandburg, 2001). The term has since come into common use. A growing amount of legal, as well as management literature, both academic and non-academic, deals with the discussion on the term, the business models of patent trolls and their impact on innovation, the patent system, and economic welfare (see section 2.3.2). Fact is that these types of companies enforce patents without having their own physical products. Patent aggregating companies do not produce physical goods.

Therefore, some business models of patent aggregating companies can also be called patent trolls because their strategy is to buy patents and enforce them (For a discussion on the definition, see section 2.3.2).

Patent trolls cause controversy, and emotions run high. On the one hand, an original patent owner who sells patents to a so-called patent troll has several advantages through this transaction: the patent owner is no longer engaged in the enforcement of the patent, they have not to bear the risks of enforcement, and they generate an immediate cash inflow. Without so-called patent trolls, many patents would not be enforced due to a lack of financial or human resources, or to lacking engineering and legal skills to detect and assert infringements. On the other hand, the infringer often has to pay high royalties, and out of court settlements with mutually beneficial cross-licensing arrangement are not possible.

Original patent owners who are interested in selling patents to patent aggregating companies that enforce infringed patents should assess the short-term benefits over the costs of a systemized patent enforcement system conducted by third parties. As a producing company, the original patent owner could be a profiteer one day and a prey at the next. So-called patent trolls make no distinction between infringing companies that are their clients or not. Therefore, selling infringed patents to a patent aggregating company that enforces the patents fuels the system and may lead to the situation that professionalized systematic patent enforcement costs the original patent owner more than the enforcement of their patents rewards him.

6.1.3 Framework for the utilization of patent aggregating companies

It has been shown that patent aggregating companies can be utilized for several value-generating options, such as selling patents with or without technologies, licensing patents to enforce them or to multiply technologies, donating patents, or selling royalties. Even though they theoretically support many value generating options, the utilization of patent aggregating companies is limited regarding their industry focus and their targeted patents. Additionally, producing companies follow diverse strategic or financial objectives in leveraging patent portfolios. Based on these three parameters, the industry and patent focus, as well as the objectives of the companies, a typology-based management framework is suggested. Figure 42 visualizes the management framework for leveraging producing companies' patent portfolios by utilizing patent aggregating companies.

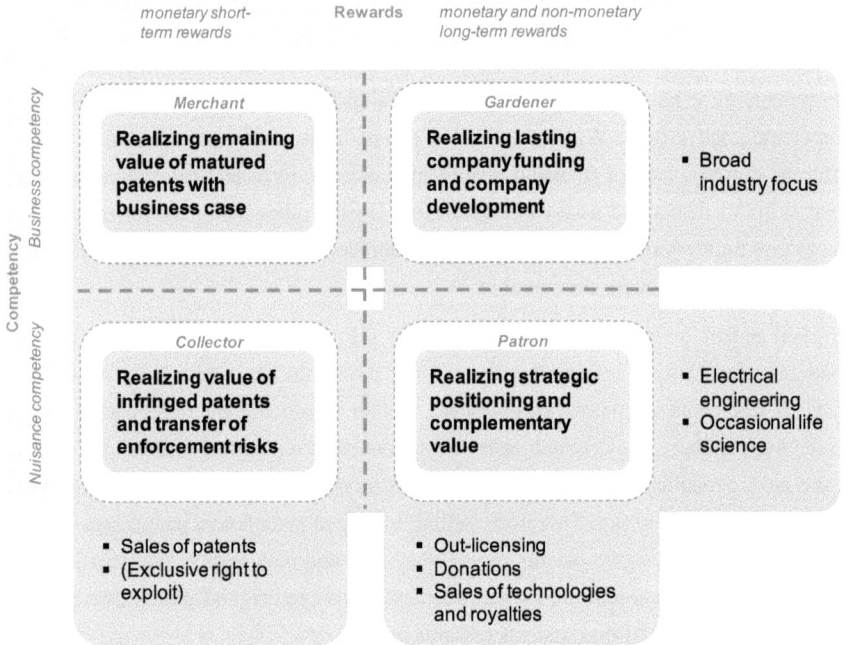

Figure 42: Management framework for utilizing patent aggregating companies

Based on the R&D strategy and the patent strategy, a producing company decides whether to exploit patents internally or externally. This decision has to be made before deciding to utilize a patent aggregating company. Therefore, the management framework assumes this strategic decision as granted and supports the operational decisions of which patents can be leveraged with the support of which archetype of patent aggregating company.

The strategic decision on how to leverage each patent is the basis for the management framework. Therefore value generating options, expectations of rewards, patent characteristics, market environment, and the company's industry are the parameters for the management framework.

Based on the systematic management of the patent portfolio, the producing company decides about the internal and/or external exploitation of a patent. In patent audits, the potential, the strategic fit, and the legal quality of the patent are analyzed. As a result, the producing company decides whether to assign or to out-license the patent.

Patents covering mature technologies already used in the market but do not fit the producing company's strategy any longer can be sold, and the original patent owner generates an immediate cash inflow. To prevent internal costs of searching for potential transaction partners, negotiating with transaction partners, and monitoring the execution of the transaction, the producing company can offer the patents to the archetype merchant. Utilizing the archetype merchant can help original patent owners to realize the remaining value of mature patents but still have a business case. Companies of all sizes, research institutions, and universities can utilize the archetype merchant.

Companies of all sizes, research institutions, and inventors from high-technology industries can also utilize the archetype collector. In addition to selling abandoned patents, infringed patents can be sold to the archetype collector. Utilizing the archetype collector can help the original patent owner to realize value from infringed patents and transfer the enforcement risk. Companies that can control the exposure of litigation lawsuits may generate more revenues from enforcing the patents without utilizing the archetype collector. For utilizing the archetype collector, the original patent owner always pays an indirect fee for the risk transfer. Discounted purchase prices reflect this indirect fee.

If the results of the patent audit show that a patent covers a technology in the core business of the company, other value generating options rather than selling may be the first choice. In addition to an internal exploitation, patents can also be leveraged externally by utilizing patent aggregating companies. Patents covering embryonic technologies or patents that already generate steady cash flows can be transferred to the archetype gardener. Utilizing the archetype gardener can help the original patent owner to realize long-term company funding and company development. Companies of all sizes and research institutions can utilize the archetype gardener. SMEs and research institutions, in particular, benefit from engaging with the archetype gardener.

Large companies and research institutions from high-technology industries can also utilize the archetype patron to leverage patents covering core technologies. Patents in areas with a high likelihood of infringement can be transferred to the archetype patron without giving up future benefits from the patent. Utilizing the archetype patron can help the original patent owner to realize a better strategic position in the product market and to realize complementary value. However, original patent owners are limited in the utilization of the archetype patron because this archetype has a limited and very selective demand and can only be used in certain times and in specific technological areas.

6.2 Development of patent aggregating companies

Patent aggregating companies are a new empirical phenomenon characterized by continuously changing business models. An analysis of past trends may help to understand the altering patent aggregating business. With a better understanding of the past, future developments and modification of the utilization of patent aggregating companies may better be estimated.

Analyzing the emergence, maturation, and utilization of patent aggregating companies three major trends can be observed (see Figure 43):

- *Trend 1*: The first business models of patent aggregating companies were interest groups; nowadays, patent aggregating companies serve as investment opportunities.
- *Trend 2*: Emergence of new patent aggregating business models is not only the action of entrepreneurs but also a reaction to existing business models.
- *Trend 3*: The assets patent aggregating companies focus on have expanded from purely patent transfer to a transfer of technology and knowledge.

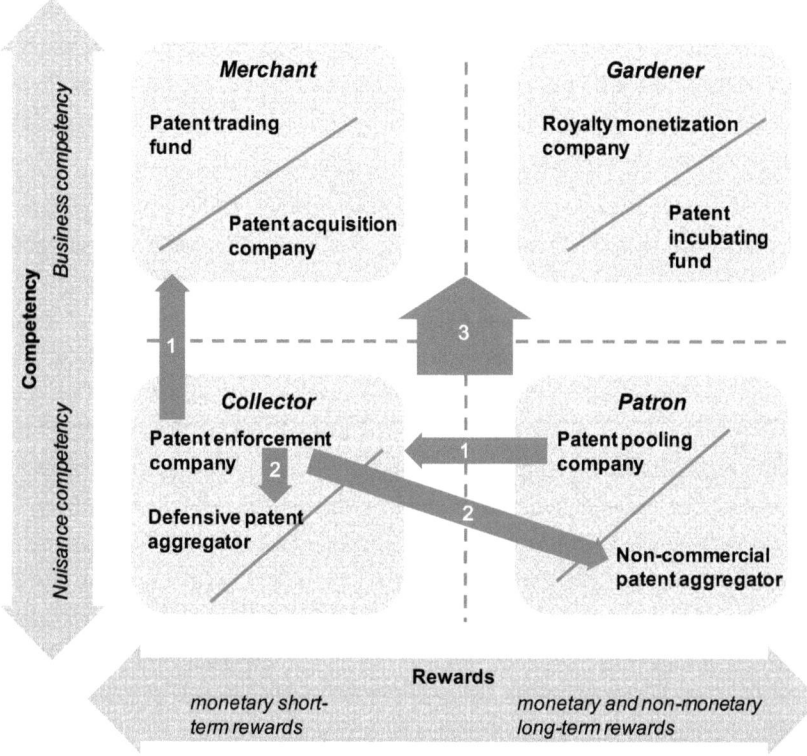

Figure 43: The three major trends that drive the evolution of patent aggregating companies

In the following sections the three trends are described in detail.

6.2.1 Trend 1: From aggregation of interest to aggregation of investments

Even though patent aggregating companies have emerged as a new phenomenon associated with the development of market for patents and technologies and open innovation, a predecessor was already formed in 1856. During the end of the 19th century, the sewing machine manufacturers *I. M. Singer Co.*, *Wheeler & Wilson Co.*, and *Grover & Baker Co.* accused each other of infringing each other's patents. The lawyer and president of *Grover & Baker Co.* Orlando B. Potter proposed to pool the respective patents, rather than to sue each other into bankruptcy. Hence, a group of joint interest was created. In addition, the sewing machine patent pool supported the manufacturer interests sustaining artificially high prices for the licensed machines[31] (Serafino, 2007). Recent patent pooling companies were set up to address standardization issues and solve the problem of patent thickets, but the joint interests of patent owners still driven them.

The first modern patent aggregating companies have emerged, taking advantage of the combination of two factors: (1) an increasingly complex patent landscape, (2) patent's transition from legal rights to company's assets and an increase in patent transaction, as well as the fact that many companies, for a long time, have not been aware of these two factors. The number of patent applications is increasing, and patent offices in the US and Europe show significant backlogs of patent applications. Therefore, it can no longer be guaranteed that all granted patents are of high quality. That leads to an increasing risk for producing companies regarding unintended patent infringements, patent thickets, and uncertainty of patent granting. Since BlackBerry maker *Research in Motion* agreed to pay USD 612.5 million to patent holding company *NTP* to settle a long-running dispute in 2006 (Magliocca, 2007), the public has been aware of the potential size of patent infringement lawsuits. Therefore, business models that take advantage of infringed patents and that are driven by entrepreneurial spirit have emerged. One interviewee stated he started his business when he met an entrepreneur in his sector because it seemed to have such a large financial potential:

> When I paid a visit to [his] office, it was like walking into Versailles. When I sat in the chair in his office, my feet did not touch the ground. I got that. I have had psychology classes and I understand what he was trying to do.

Especially during the turn of the century, entrepreneurs took the chance and founded patent aggregating companies. Five out of six patent aggregating companies, created by daring entrepreneurs, were founded in this time.

As the market for technology and patents has evolved, the risks related with these assets and transactions are easier to evaluate, and large patent suits have come into the public eye, patent aggregating companies have moved on from entrepreneurial driven companies to investment vehicles of the financial industry. Large financial resources

---

[31] On the day the last patent expired, *I.M. Singer* reduced the price of its sewing machines by 50% in order to compete in an open market.

from institutional or private investors back 12 of the 27 sample firms, with all of them, except three, founded after 2004.

6.2.2 Trend 2: Responses to organized patent enforcement

Business models of patent aggregating companies have not only emerged because single persons or entities recognized the business opportunity. Especially during the last five years, business models have emerged as reaction to already existing, revenue generating business models and their use of patents.

Patent enforcement companies aggregate patents to enforce them and generate revenues through a stick licensing approach. In patent infringement lawsuits between producing companies, often both companies use their own patent portfolios as basis for negotiation. Agreements are often closed not only through licensing payments but also through cross-licensing agreements. Patent enforcement companies are not interested in cross-licensing agreements, and defendants have to pay full licensing fees. As a reaction to this business model, defensive patent aggregators have emerged. Many patents that are interesting for patent enforcement companies are freely available in the market. A single producing company is not able to buy all relevant patents to prevent a patent enforcement company from getting hold of them. Therefore, defensive patent aggregators bundle interests of several companies and acquire potential threatening patents. This action prevents infringement lawsuits against producing companies. Therefore, the business model of defensive patent aggregators is only a reaction to the business model of the patent enforcement company and does not exist without it.

As patents are used to block competitors in certain areas, this behavior not only impacts the revenue level of firms but in certain cases, also impacts the wellbeing of the society or groups of people. For instance, to be forced to pay licensing fees, users in less developed countries are excluded from certain patented technologies. In addition, certain technologies that foster sustainability and resource saving technologies face patent thickets or patent holdings of patent enforcement companies. This situation prevents innovation and technological progress. Therefore, as reaction to blocking positions and patent enforcement strategies of other patent aggregating companies (as well as producing companies), the business model of non-commercial patent aggregators has evolved. Patents held by a non-commercial patent aggregator do not exclude users from technology. Non-commercial patent aggregators offer, mostly free of charge, access to technology and could be able to foster innovation or improve the conditions of disadvantaged groups.

6.2.3 Trend 3: From enforcement agents to innovation intermediaries

Patent aggregating business started with a focus on aggregating the legal right of exclusion. The first business models are based on the exclusion of third parties and the enforcement of patents (patent enforcement companies, patent pooling companies). As the innovation paradigm has changed from closed innovation to open innovation, the transfer of patents and technologies has become important for a firm's innovation process. This change is also observable in the change of the patent aggregating companies' business models.

According to Chesbrough (2006), innovation intermediaries help technology providers to find buyers or licensees for their technologies. In return, innovation intermediaries allow technology buyers to use technologies they do not have developed in a rapid and beneficial fashion. Innovation intermediaries occur in the two major forms of agents and brokers (Chesbrough, 2006):

- *Agents*: These companies represent only one side of the technology transaction.
- *Brokers (or market makers)*: These companies match buyers and sellers of a technology, shape the terms of the transaction, and sometimes support the commercialization of technologies.

Analyzing the activities of the archetype merchant, as well as the archetype gardener (or more specifically patent incubating funds because royalty monetization companies base their business models on existing licensing agreements) leads to the conclusion that these archetypes are innovation intermediaries (according to Chesbrough, 2006). While aggregating patents from original patent owners, these archetypes help firms to exploit their technologies externally. Business models of both archetypes aggregate patents and sell or out-license them to generate revenues. Therefore, they match the supply and demand of technologies; hence, they are innovation intermediaries, or more specifically, brokers or match makers (according to Chesbrough's (2006) definition).

Patent aggregating companies are no longer the sole buyer of enforceable rights but have developed into transaction parties and transaction enablers in the market for technologies.

6.3 Meeting demand for learning effect as driving factor

As markets for patents and technologies lack transparency and are characterized by high transaction costs, third parties as enabler of transactions have emerged. Defensive patent aggregators, patent acquisition companies, patent enforcement companies, patent incubating funds, and patent trading funds acquire patents and fulfill a match making function in a very broad sense; therefore, they can be seen as intermediaries in the market for patents and technologies. The emergence and the activities of intermediaries can be discussed from the perspectives of different economic theories. From a transaction costs economics perspective (Williamson, 1975, Williamson, 1985), intermediaries can help buyers and sellers set up appropriate governance mechanisms for executing risky transactions (Benassi & Di Minin, 2009). From a network theory perspective (Burt, 1995), Burt (2005) suggests that intermediaries emerge as a function of structural holes in a network structure (as cited in Benassi et al., 2010).

The following part goes beyond explaining the emergence of intermediaries in the market for patents and technologies. It discusses the development of patent aggregating companies from enforcement agents to innovation intermediaries. Drawing on a resource-based perspective (Barney, 1991; Lavie, 2006; Wernerfelt, 1984), the main driver for the trend that patent aggregating companies develop to innovation intermediaries can be found with the original patent owners and their demand for learning effects to adapt to the changing environment. Learning effects are non-monetary benefits the patent owner can realize by transferring patents to the patent aggregating company. The total benefits can be enunciated in a mathematical expression that helps to explain the observed trend in a more formal way. Therefore, the first two parts illustrate the elements of the benefit function. The subsequent part explains the trend and the main driver.

6.3.1 Monetary benefits of utilizing patent aggregating companies

In Chapter 4, potentials offered by patent aggregating companies are illustrated. These potentials result in both monetary (B_M) and non-monetary benefits (B_N) for the original patent owner. By assigning patents to a patent aggregating company, the original patent owner can realize the total benefits (B_T) resulting from monetary and non-monetary benefits.

Monetary benefits are the pure financial compensation an original patent owner receives in patent transactions. They result from the potential for resource enhancement and access to financial resources offered by the patent aggregating companies. For transferring the ownership rights to a patent aggregating company, the original patent owner receives a lump sum payment (LP). A lump sum payment is a single payment for a patent paid by the patent aggregating company. The original patent owner receives an immediate cash flow, and future payments are not made. Another form of compensation is an upfront payment (UF). An upfront payment is an amount of money delivered at the time the contract is signed. Additionally, other types of payments are made during the lifetime of the contract. In licensing agreements, royalties are usage-based payments made by the licensee to the licensor for the right to use a patent. Typically, the amount of the royalty payments is dependent on a percentage of gross or net revenues derived from the use of the patent or a fixed price per unit sold of an item. Royalties are paid over a certain period of time and depend on the time the patent is used. If the original patent owner transfers the patent to a patent aggregating company, which out-licenses the transferred patents, the two parties can agree to share the royalties, and the original patent owner receives partial royalties (PR). Some patent aggregating companies mandate the original patent owner to advance the transacted technology. In this case, the original patent owner receives compensation for its R&D efforts (RD). Additional direct monetary benefits, however, not directly paid in cash by the transaction partner, are savings of patent maintenance costs or other directly related costs (CS). Transferring the ownership of patents is directly related to costs savings for the original patent owner because the patent aggregating company then covers renewal fees, enforcement fees, and other maintenance costs. Donating patents to non-profit organizations, the original owner is allowed to claim tax deductions. This tax deduction directly affects the profits of a producing company and results in tax savings (TS).

The actual compensation or compensation bundles depend on the contractual agreements between the parties. However, the monetary benefits B_M are a function of above-mentioned factors:

$$B_M = f(LP, UF, PR, RD, CS, TS) \tag{1}$$

6.3.2 Non-monetary benefits of utilizing patent aggregating companies

Besides monetary benefits, the original patent owner can also realize non-monetary benefits. These non-monetary benefits also result from the potentials patent aggregating companies offer. They include all additional benefits that do not directly influence the financial balance sheet of the original patent owner.

One non-monetary benefit is the transfer of risks (TR). Patent aggregating companies offer producing companies the potential for risks reduction. Original patent owners can benefit from the transfer of two types of risks: (i) the risks that result directly from R&D activities, for instance, developing commercially unsuccessful inventions or loss of royalty streams resulting from licensing agreements based on own R&D results; and (ii) the risks that result from enforcing the patented R&D results, for instance, the costs of patent infringement lawsuits or damaged reputation.

Patent aggregating companies offer the external potential for market fostering. As an active buyer in the market for patents, they create demand for patents and offer liquidity. In addition, new business models create demand. Patent aggregating companies offer potential for market interaction and support producing companies in identifying opportunities within companies' patent portfolios. These potentials create new or additional leveraging options for the original patent owner that then benefits from a broader spectrum of opportunities (OA). Assigning patents to patent aggregating companies opens up the opportunity for the original patent owner to invest in R&D, innovation processes, or commercialization activities. All these activities support the strategic position of the original patent owner.

Potentials for market interaction and patent aggregating companies as the buyer have resulted in an additional benefit for the patent owner. In conventional patent transactions between the original patent owner and a potential patent buyer, both parties are known. Patent offerings in the open market disclose information about changes in patent and technology strategies. If the identity of the patent owner is known, this can also influence the price. Selling patents to a patent aggregating company results in the non-monetary benefit of anonymity while satisfying the demand for patents for other producing companies (AD).

Original patent owners often face personal constraints that lead to underutilized market or technological opportunities. Patent aggregating companies offer potentials to resource enhancement. Realizing these potentials, the original patent owner can benefit through additional human resources (HR), but also through additional or complementary competencies (CO).

Licensing agreements always carry the risk of loss of royalties due to terminated products or insolvent licensees. Transferring patents to patent aggregating companies not only has the advantage of immediate cash inflows (monetary benefit) but the original patent owner can also benefit from planning security (PS) and an opportunity to extend the strategic scope.

A general benefit is the value of the grant back license (GB). The value of a grant back license results in the inflow of new ideas and innovations; hence, the improvements or innovations made by new patent owner or licensee have to be transferred to the original patent owners. That can improve the own technological position without having additional R&D expenditures.

Patent aggregating companies offer potentials for cost effectiveness and have competencies in the market for patents. Collaborating with patent aggregating companies gives the original patent owners the non-monetary benefit of learning effects (OL). A close collaboration between patent aggregating companies and original patent owners also offers learning potential for competencies regarding R&D or commercialization of innovations.

In summary, the monetary benefits B_N are a function of above-mentioned factors:

$$B_N = f(TR, OA, AD, HR, CO, PS, GB, OL) \tag{2}$$

Hence, the total benefit a patent owner can generate by assigning patents to a patent aggregating company is:

$$B_T = B_M + B_N \tag{3}$$

6.3.3 Benefits depend on the type of patent aggregating company

As the trend from the function of transferring sole legal rights to an innovation intermediary function is discussed in this part, the analysis focuses on the patent aggregating companies that acquire patents and fulfill a match making function in a very broad sense: defensive patent aggregators, patent acquisition companies, patent enforcement companies, patent incubating funds, and patent trading funds.

Figure 44 depicts the non-monetary benefits (B_N) and the monetary benefits (B_M) for original patent owners that assign a patent to a patent aggregating company. B_T shows the total benefit for a patent owner resulting from this transaction. The relative height and shape of the curves are only rough estimates because they also depend on a number of secondary factors and company specific characteristics. The mathematical expression of the benefits for the original patent owners is an attempt to explain the observed trends by clarifying the situation a patent owner faces and a patent aggregating company provides in a more formal way.

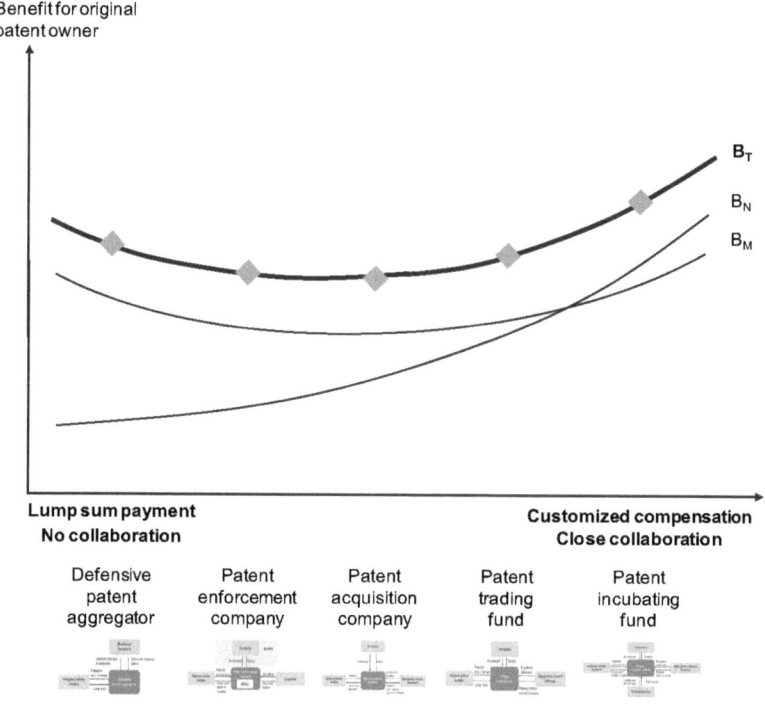

Figure 44: Resulting benefits for the original patent owner

An empirical trend towards patent aggregating companies as innovation intermediaries is observable. The changing strategies of the original patent owners, as well as the changing market conditions are able to explain this trend. In patent transactions, the original patent owner no longer focuses on sole revenue generation by selling useless patents but increasingly becomes aware of the diverse benefits patents offer. The total benefit curve reflects this development.

The curve representing the monetary benefits (B_M) is U-shaped. Defensive patent aggregators and patent enforcement companies buy only infringed patents and have the potential for large revenues from stick licensing. Therefore, these patent aggregating companies pay relatively high prices, mainly as lump sum payments, for the patents. As defensive patent aggregators act as insurance for the attached companies and can only operate successfully when they get hold of the targeted patents, they may pay even a higher price than patent enforcement companies may. In addition, patent enforcement companies may display their experience and market power and not offer the market price to companies. Patent acquisition companies buy all sorts of patents and on average; they pay only low prices for the patents. Only in very few cases, agreements on compensation bundles including other payments than a lump sum payment are signed. As patent acquisition companies buy a broad range of patents, it is possible that original patent owners will be able to sell patents they could not exploit otherwise. Patent trading funds buy patents covering already commercialized technologies and exploit them either through selling or through out-licensing agreements. As technologies are older and easier to evaluate, the price range for the acquired patents is limited but often higher than prices paid by patent acquisition companies as the technical applicability is already shown. In cases of licensing agreement, royalties are paid over a longer period of time and therefore, have to be discounted to estimate the monetary benefit. Therefore, the B_M-curve does not show large differences between these two business models. Patent incubating funds aggregate patents covering embryonic technologies, invest in the advancement of patents and technology, and exploit the technology through a carrot licensing approach. The compensation or compensation bundle paid by the patent incubating fund is customized to the situation of the original patent owner. Often patent incubating funds pay upfront payments. If the technology is out-licensed, the original patent owner participates on the royalties. As the original patent owner has important tacit knowledge of the technology, patent incubating funds mandate them for further development. That leads to additional cash inflows from the R&D contract. The original patent owner receives royalties and payments for R&D efforts over a period of time. Due to the time factor, the payments have to be discounted, and the curve of monetary benefits increases only slightly.

The curve representing the non-monetary benefits (B_N) increases from the left hand side of the horizontal axis representing points without collaboration between patent aggregating company and original patent owner and fixed payment systems to the right hand side, representing close collaboration between the parties and customized compensation. All patent aggregating companies offer some elements of the B_N-function, such as risks transfer. However, defensive patent aggregators and patent enforcement companies offer the smallest non-monetary benefit to the original patent owner. Defensive patent aggregators may deliver a reputation and marketing tool for not being involved with unpopular companies but this non-monetary benefit seems to be smaller than some non-monetary benefit resulting from anonymity or competency enhancement offered by patent enforcement companies. Compared with defensive patent aggregators and patent enforcement companies, the collaboration with patent acquisition companies, patent trading funds, and patent incubating funds offers larger non-monetary benefits. Besides a hedging opportunity for R&D risks, they enhance human and financial resources and support companies in the identification of opportunities. The major difference between all business models also explains the increase in the B_N-curve and hence, the trend from enforcement agents towards innovation intermediaries is the benefit of learning effects.

According to organizational learning theory (Fiol & Lyles, 1985; Huber, 1991; Levitt & March, 1988), companies have to recognize the changing environment and change their goals and actions to stay competitive. As the market for patents and technologies has emerged and the innovation paradigm has changed to a more open approach, external patent exploitation has become more important. Companies have to adapt to these changes in the environment. For producing companies, external patent exploitation and the transfer of technology and knowledge is not a core business (Davis & Harrison, 2001; Lichtenthaler & Ernst, 2009). Often, these transactions are conducted in ad hoc projects. Dedicated resources do not exist and therefore, external patent and technology transactions are often unsuccessful (Lichtenthaler, 2011; Rivette & Kline, 2000). According to the resource-based view, corporations can create a competitive advantage through the development and intelligent application of core resources and capabilities (Barney, 1991; Grant, 1996; Wernerfelt, 1984). Therefore, by adopting a resource-based perspective on the original patent owner, the lack of internal competencies leads to high transaction costs (Cohen & Levinthal, 1990; Lane et al., 2006). Firms may influence their transaction costs by developing internal competencies based on learning effects (Kale et al., 2002; Lichtenthaler & Lichtenthaler, 2009; Silverman, 1999). Therefore, to adapt to emerging markets for patents and technologies and to exploit patents and technologies externally and optimally, companies take actions and increase their demand for learning effects regarding innovation transfer. Patent acquisition companies, patent trading funds, and patent incubating funds transfer patents and technologies. According to Chesbrough (2006), they are an innovation intermediary.

In analyzing the five patent aggregating companies, it becomes apparent that the benefit of the learning effects they offer increases from the defensive patent aggregator, which does not offer any collaboration, to patent incubating funds, which offer close collaboration. Defensive patent aggregators and patent enforcement companies can be seen as patent intermediaries but not as innovation intermediaries. In addition, the learning effect from their patent transaction is very limited. As defensive patent aggregators often detect and buy interesting patents via patent brokers, the original patent owner does not interact with the defensive patent aggregator. Therefore, they are not able to benefit from learning effects. Patent enforcement companies may offer marginal learning effects regarding negotiations or patent enforcement. Patent acquisition companies are innovation intermediaries because they also transfer technologies. However, they interact with the original patent owner only to a limited extent and therefore, offer only limited benefits from learning effects. The potential learning effects are mainly in the area of patent auditing or opportunity identification. Patent trading funds collaborate with the original patent owner to a certain degree. Consequently, the original patent owner can realize learning effects regarding patent valuation, patent management, and applying technologies in other industries. Patent incubating funds offer the largest learning effects. This is represented by the graphical intersection of the B_N- and the B_M-curve that indicates that the offered non-monetary benefits of patent incubating funds excel the sole monetary benefits. Patent incubating funds advance the acquired technology in collaboration with the original patent owner. Therefore, the original patent owner can realize learning effects regarding R&D, commercialization of technologies and patents, marketing, patent management, and patent exploitation competencies.

The discussion shows that with respect to the general economic environment of changing innovation paradigm and maturing markets for patents and technologies, revenue generation is only one part of a patent transaction. Non-monetary benefits become increasingly important. In particular, the objective of patent owners to establish their competencies of external patent exploitation and technology transfer has become an important factor. In collaborations with third parties, original patent owners can benefit from the experience and competencies of the partner and realize learning effects to establish internal competencies. Recognizing this prospect, original patent owners ask for learning effects from innovation intermediaries. The trend of patent aggregating business models from enforcement agents to innovation intermediaries reflects this behavior.

6.4 Summary

Patent aggregating companies can be utilized for several value creating options. Even though a promising alternative for leveraging patent portfolios, the utilization of patent aggregating companies is not free of constraints. The industry, in which the producing company operates and hence, patents, is the basis for all opportunities to cooperate with a patent aggregating company. The value of the patents and the technology phase of the life cycle are important constraints that narrow down the selection of patent aggregating companies. In addition, macro-economic impacts on the patent system have to be considered. In sum, these constraints and the typology developed in Chapter 5 lead to a management framework that is able to show the patent managers of producing companies which archetype of patent aggregating company is suitable for which patent portfolio leveraging activity. Therefore, it answers the research question: *How are patent aggregating companies utilized to leverage patent portfolios of producing companies?*

As patent aggregating companies operate in a fast changing environment, not only is the status quo important for the patent manager who seeks answers to the question whether patent aggregating companies are an option for producing companies but also the development and direction these business models head in is important. During the last two decades, three major trends of patent aggregating companies' development could be observed: (1) Patent aggregating companies have developed from interest groups to investment vehicle; (2) patent aggregating companies have emerged as a reaction to existing business models; and (3) patent aggregating companies have evolved from enforcement agents to innovation intermediaries. Reflecting the overall research question: *Patent aggregating companies are an option for producing companies?*, the last trend, in particular, is important for patent managers. As producing companies now have to exploit their patents externally, and as part of the open innovation paradigm have to trade technologies and knowledge to stay competitive, it is important to build internal resources. Companies are aware of this requirement and demand cooperation partners from whom they can learn how to leverage the patent portfolio externally. Satisfying this demand, the business models of patent aggregating companies have emerged from companies that buy infringed patents without cooperating with the original patent owner to companies that trade patents and technology and work closely with the original patent owner.

7 Conclusion

With the objective to analyze the recent phenomenon of companies that do not have R&D or produce physical goods but buy patents on a large scale, this study analyzes 27 patent aggregating companies from the US and Europe using an exploratory research design. Based on the analyses and discussions of the previous chapters, the following chapter summarizes the key findings, highlights the central contributions for management theory and practice, and looks ahead to further research and trends.

7.1 Contribution to management theory

This thesis constitutes the first study of patent aggregating companies that goes beyond examining descriptive issues based on the narrow focus of technology market intermediaries or NPE. In the light of increasing interest in technology licensing and external patent exploitation (Rivette & Kline, 2000) and markets for patents and technologies (Arora & Gambardella, 2010a), it also constitutes the first study on different motives, business models, and the development of patent aggregating companies, as these companies fulfill a special intermediate function between patent supply and patent demand. Thus, it contributes to overcoming the low emphasis on empirical research into technology market intermediaries and external patent exploitation, which has only recently been highlighted (Arora & Gambardella, 2010a; Howells, 2006; Lichtenthaler & Ernst, 2008a; Tietze, 2011). This study provides the following specific contributions:

Contribution 1: Clarification of the term patent aggregating companies

This study contributes to literature on markets for technology, external patent exploitation, and technology market intermediaries by investigating and confirming the new phenomenon of patent aggregating companies, defining this phenomenon, and clarifying the strategies and business models of this new phenomenon.

The literature on the market for patents and technologies focuses mainly on producing companies as seller and buyer, problems of producing companies, and technology market intermediaries as facilitator of transactions. Publications that recognize buyers that do not produce goods focus on companies that acquire only infringed patents to enforce them, so called NPE. Therefore, the term patent aggregating company is often used synonym with the definition of NPE. Only scattered publications recognize patent aggregating companies with other motives, but they are limited to offensive and defensive buying motives. In general, literature on patent management that analyzes the reasons why firms patent or why companies acquire patents focuses on producing companies and their patent strategies. Current research neglects companies that do not have products, and therefore, may have different reasons to acquire patents. Only in the context of NPE, publications mention that these types of companies acquire patents to enforce them.

This study closes this gap in literature and finds that there is no single dominant motive for patent aggregating companies. Rather, patent aggregating companies can be grouped into eight different business models (defensive patent aggregator; non-commercial patent aggregator; patent acquisition company; patent enforcement company; patent incubating fund; patent trading fund; patent pooling company; royalty monetization company) according to their specific motive to aggregate patents. Based on these results, the study shows that patent aggregating companies are not equatable with NPE. To underline this finding, a definition of patent aggregating companies is derived. This definition helps to distinguish them from other patent service providers, patent intermediaries, and NPE. In addition, the detection of the different reasons to aggregate patents goes beyond the conventional focus of why producing companies patent or acquire patents. Thus, analyzing the reasons why patent aggregating companies acquire patents enhances our understanding in patenting motives.

The definition and the different business models of patent aggregating companies extend the understanding of the new players in the market for patents and technologies and may serve as a base for future research.

Contribution 2: Conceptualizing patent aggregating companies and deriving a typology of patent aggregating companies

This study contributes to literature on patent intermediaries and patent management by investigating activities of patent aggregating companies and the services and benefits they offer patent owners. Based on the results, four archetypes of patent aggregating companies – the merchant, the gardener, the collector, and the patron – are identified.

Literature on technology market intermediaries investigates the activities intermediaries conduct to match supply and demand. Most publications focus on the single tasks the intermediaries perform in the transaction process (e.g., Benassi & Di

Minin, 2009; Howells, 2006; Lopez-Vega, 2009; van Lente, Hekkert, Smits, & van Waveren, 2003). A delineation of patent aggregating companies or a description of their activities is lacking. Publications are limited to descriptive issues and often fail to systemize technology market intermediaries. Only Benassi and Di Minin (2009) attempt to derive a taxonomy of patent brokerage, which includes seven different patent brokers, two of them patent aggregating companies according to the definition proposed in this study. A further conceptual clarification of patent aggregating companies is lacking. Especially in the light of the difficulties producing companies face to leverage their patent portfolios optimally, a conceptualization of companies to support producing companies is missing. Extant literature describes a lack of transparency, asymmetric information, and high transaction costs in the market for patents and technologies, but previous researchers have only identified the problems companies face but are not able to provide solutions to these problems.

This study addresses this blank spot in literature and identifies four archetypes of patent aggregating companies: the merchant, the gardener, the collector, and the patron. These archetypes differ significantly regarding their competencies, the rewards they offer the original patent owner, and the breath of transaction they focus on. The results are of special interest for patent managers of producing companies that seek support for leveraging the company's patent portfolio. The conceptualization in four archetypes allows the manager to identify which archetype would best suit the strategic objective of the patent portfolio's leveraging activities and hence, could be utilized by the producing company. The typology may be applied in a descriptive way to analyze patent aggregating companies further. Additionally, it may be used in a normative way to develop patent portfolio leveraging strategies that include or purposely exclude patent aggregating companies. Thus, this typology may help managers pursue a more systematic patent portfolio leveraging approach, and as a conceptualization of patent aggregating companies, it may serve as a basis for future research.

Contribution 3: Detection and explanation of trends in patent aggregating companies' business models

This study contributes to literature on technology market intermediaries, which includes publications on innovation intermediaries, by analyzing the development of patent aggregating companies and the driving factors behind the major trends. Drawing on a resource-based perspective, the demand of producing companies for learning effects has forced patent aggregating companies to evolve from enforcement agents to innovation intermediaries.

Authors that analyze the emergence of technology market intermediaries are scarce. Information asymmetries in the market for patents and technologies are the main explanation why technology market intermediaries exist. Drawing on transaction costs economics theory or network theory; attempts are made to clarify the question of existence (e.g., Benassi & Di Minin, 2009). Even though the market for patents and technologies is a constantly changing market and technologies market intermediaries emerge and vanish fast (Benassi & Di Minin, 2009; Millien & Laurie, 2008), previous research has not investigated why firms and entrepreneurs have moved on and now follow other activities and business models and offer other services to the original patent owners.

This study responds to this shortcoming in literature, analyzes trends and changes in the business models of the 27 case firms, and reveals three major trends in the development of patent aggregating companies. Overall, a trend that patent aggregating companies shift the focus from amassing infringed patents to transferring patents, technologies, and knowledge is noticeable. According to Chesbrough (2006), patent aggregating companies have developed from enforcement agents to innovation intermediaries. Integrating organizational learning theory and resource-based view of the firm, this trend appears to result from the fact that original patent owners seek to establish their own competencies of external patent exploitation and technology transfer. Original patent owners no longer focus on sole revenue generation by selling useless patents, but increasingly have become aware of the non-monetary benefits from these transactions and the large learning potential innovation intermediaries offer. To meet changing economic conditions, companies collaborate with experienced partners. That enables companies to benefit from learning effects. Thus, original patent owners have increased the demand for learning leading to the emergence of patent aggregating companies as innovation intermediaries. The explanation of the trend towards innovation intermediaries contributes to the discussion on innovation intermediaries and their functions and broadens the application of the term 'innovation intermediary'. The trends of patent aggregating companies' development further lay emphasis on the changes and the transitory nature of business models in the market for patents.

7.2 Implications for management practice

Patent aggregating companies have emerged as a recent, not yet well-understood phenomenon. As patent aggregating companies are significant players in the market for patents and technologies, they could help to overcome producing companies' impediments to external patent exploitation and to leverage producing companies' patent portfolios optimally. Therefore, this research on patent aggregating companies gives some insights into the benefits patent aggregating companies might give producing companies.

The insights gained in this study show the different strategies, motives, potentials, and activities of patent aggregating companies. Based on these insights that are reflected on literature, recommendations for patent managers of producing companies can be drawn. Therefore, the managerial recommendations provided in the following part refer to leveraging patent portfolios of producing companies and to what patent managers should consider by utilizing patent aggregating companies.

Recommendations to utilize patent aggregating companies for leveraging patent portfolios

Patent aggregating companies differ substantially regarding their business models, their competencies, and the rewards they give to original patent owners. Therefore, a patent manager cannot utilize every patent aggregating company for every value generating option of patent exploitation. The management framework developed in section 6.1.3 provides a guideline to select the suitable archetype of patent aggregating company. In addition, to utilize patent aggregating companies optimally, the following recommendations for patent managers in producing companies are provided:

Detach from the picture of the patent troll. Patent aggregating companies offer a wide range of utilization opportunities beyond the traditional enforcement agent model. Even though some patent aggregating companies buy patents to enforce them, many other patent aggregating companies offer other benefits and potentials for patent portfolio leveraging activities. It is important that patent managers of producing companies see the wide range of patent aggregating companies and are open to using it. In some cases, patent enforcement companies are opponents in infringement lawsuits but knowing who these companies are and what they do helps managers to react to them. Even though patent aggregating companies and the original patent owner might be opponents in some cases, patent aggregating companies could take over the enforcement risks in other cases. A critical examination is more beneficial than a general damnation.

Before selecting a patent aggregating company, define the initial position of the producing company. It is important that a company that plans to cooperate with a patent aggregating company is aware of its own resources and competencies, and of the inventory of its patent portfolio. To choose a suitable patent aggregating company and to utilize it optimally, the patent manager has to analyze its resources and evaluate its competencies. Only with this analysis, can the patent manager choose a patent aggregating company that has additional or complementary resources and competencies. Employing patent aggregating companies with similar resources and competencies does not enhance the leveraging position and costs money without benefiting from it because the producing company could conduct the offered services without support. For the company, it is also important to know the inventory of the patent portfolio and the value of the patents. Offering a whole, not-preselected, patent portfolio to a patent aggregating company could result in three major disadvantages for the producing company: (1) spending time and resources for communication and collaboration with the patent aggregating company with uncertain return; (2) potentially revealing patenting strategies or innovation strategies by disclosing all relevant information for evaluation; (3) choosing a patent aggregating company that is not suitable for the type of their patents that should be exploited. Evaluating the patents initially helps to save costs, time, and resources for the producing company. A thorough evaluation also prevents failings in external patent exploitation due to uncovering patents with little value in the first place. Only in cases were patent aggregating companies are employed for patent portfolio audits, is it justifiable that the producing company hands over entire unevaluated portfolio.

Define the objectives that should be achieved by utilizing a patent aggregating company. As patent aggregating companies offer different competencies and benefits, it is important that the original patent owner define objectives regarding the financial return, the intended organizational learning, and the relationship between the patent aggregating company and the original patent owner. Based on this objective function, the patent manager can select a patent aggregating company that is able to achieve the objectives. In particular, the financial return and the relationship have to be aligned with the results from the analysis of the initial position to prevent a misjudgment of the situation and a derivation of unrealistic objectives. It is advisable that the patent manager reflects the objective function on the patenting and innovation strategy. To realize the full potential of external patent exploitation projects, the bigger context of open innovation and as a next or combined step, the possible innovation acquisition should be considered.

Understand the limitation of patent aggregation. Even with a tempting opportunity for patent managers to give away unused patents, the business models of patent aggregating companies do not fulfill a broad-spectrum function. After determining the position, crafting the objectives, and identifying valuable patents, the patent manager can select the suitable patent aggregating company based on the results of these activities. If the evaluation shows only patents with low values, it is not advisable for the patent manager to spend resources contacting patent aggregating companies. In addition, the stage in the product life cycle or other patent inherent factors could prevent cooperation with patent aggregating companies. Another limitation is the industry focus of patent aggregating companies. Operating in certain industries hampers cooperation with patent aggregating companies per se.

Focus on outcome but at the same time pursue objectives. To use patent aggregating companies optimally, a patent manager should not only focus on the mere result of the patent transaction but also try to utilize all offered potentials fully. If the patent manager's main objective of the collaboration is to benefit from the experience of a patent aggregating company and to build up internal resources, he/she has to provide necessary resources that are able to learn from the actions and after this, be able to transfer the learning within the company. Even if the outcome is only limited satisfactorily, the learning effect generates non-monetary benefits. Patent managers should also utilize the network patent aggregating companies offer. Future transactions or other company activities can directly benefit from the cooperation with a patent aggregating company.

7.3 Further research and trends

As the first comprehensive study, this research explores the phenomenon of patent aggregating companies and of how they can be utilized in producing companies' patent portfolio leveraging activities. Based on a rich empirical data set, the study answers the specified research questions. The following comments encourage further research into patent aggregating companies in order to gain deeper insights into these types of companies and to overcome producing companies' impediments to the market for patents and technologies.

- This research analyzes the potentials patent aggregating companies offer and describes the monetary and non-monetary benefits these companies can realize. The research applies a qualitative approach to answer how and why questions. It would be worthwhile to investigate the performance of the patent aggregating companies and the success of their business models, for example, the actual commercialization outcomes of patent aggregating companies that aggregate patent to generate revenues. An analysis of quantifiable results would help to distinguish between successful and unsuccessful business models. Only with actual performance data, is it possible to evaluate the quality and superiority of patent aggregating companies regarding patent identification, selection, enhancement, and exploitation.
- Due to the activities in the market for patents and technologies that mainly take place in the US or Europe, the selection of the empirical sample in this study focuses on companies with headquarters on these two continents. Since many production sides are located in Asia, and as China is now the country with the largest number of patent application per year, research on patent aggregating companies with headquarters in Asia could give insights into the developments and business models there. A map of Asian patent aggregating companies could show the differences between the continents, as well as give producing companies a guideline how to interact with them.
- The original patent owner and its problems to leverage the patent portfolio optimally is the drawn perspective for investigating patent aggregating companies in this study. However, this perspective is only the supply side of the patent aggregation process and therefore, only one part of the process. Equally interesting is the demand side of the patent aggregating company. New results on how patent aggregating companies can benefit or harm potential demanders of patents and on how they interact could help producing companies to optimize patent management, potential defense mechanism, and the acquisition of innovation.
- The original patent owner has not yet been a unit of analysis in the context of patent aggregating companies. An interesting path for further research is exploratory and explanatory studies on the original patent owner. Especially interesting are interaction patterns with the patent aggregating company, as well as

an assessment of the actual external patent exploitation performance of projects in collaboration with patent aggregating companies. The derived results could give additional information on the actual monetary and non-monetary benefits of cooperating with patent aggregating companies. Also further guidelines for establishing a patent exploitation process that incorporates patent aggregating companies could optimize patent portfolio leveraging activities.

- Patent aggregating companies are perceived as significant buyers in the market for patents and technologies. So far, scholars have not yet studied their actual impact on the market for patents and technologies. An assessment of the buying power of patent aggregating companies could contribute significantly to the scattered literature on the size of the market for patents and technologies. In addition, information on the size of the market could enhance the confidence of other players in the market and increase activities and with this efficiency.

Patent aggregating companies are active in a highly dynamic environment. The market for patents and technologies is far from functioning well and offers large financial potentials for new business models, entrepreneurs, and ideas. However, these large potentials are associated with high risks and the danger of failing. New business models emerge, some established business models are increasingly successful; but at the same time, many business models or patent aggregating companies vanish after only a few years of operation. In the future, the two basic reasons – enforcement of patents and innovation transfer – to buy patents will sharpen further the different business models of patent aggregating companies. Nuisance competency and business competency will be the driving factors for more specialized business models. Which of the two reasons is going to dominate the future landscape of patent aggregating companies is strongly dependent on the success of the patent aggregating companies that act as innovation intermediaries during the next three to five years.

Most of the business models that transfer innovation, such as patent incubating funds or patent trading funds, have not yet reached their exploitation and commercialization phase. Therefore, they have not shown the long-term value they might add to the economic environment. Depending on the success of these companies, the interest of financial investors will grow further. Already today, large financial resources often back patent aggregating companies. As patent aggregating companies prove the sustainability of their business models, financial institutions will broaden their investment spectrum for all types of investors. In addition to the success of the business models, external factors will drive future investment opportunities, such as the establishment of a financial market for IPR in Europe. Based on sustainable business models of patent aggregating companies and a financial market for IPR with standardized contracts and patent aggregating companies as reliable and experienced business partners, the market for patents and technologies could further develop, and investments could be made on a stable basis. However, it will be important that the invested funds do not fuel a system that is based on existing patents but that the invested capital is used to finance the creation of patents, leading to R&D and further innovation.

Whereas today, most patent aggregating companies acquire patents from all types of original owners, the acquisition activities need to focus more on particular patent owners and enforcement agents and on how innovation intermediaries might divide their source of patents. It seems that patent enforcement agents might focus more on MNEs and start to establish partnering agreements for all litigation activities with large players. This trend seems inevitable as on the one hand, large companies are becoming increasingly involved in resource intensive law suits that are not only costly but also damaging to their reputation. On the other hand, enforcement agents are seeking for business risks' decreasing opportunities and prefer to partner with companies that have large and strong patent portfolios rather than acquiring single patents from a variety of patent owners. Based on governmental funding (e.g., French Brevet, a French funding scheme that helps to create a market place for technologies of SMEs and single inventors, or the technology trading agencies that are established to transfer university inventions to producing companies), patent aggregating companies that transfer innovation from SMEs to MNEs will burgeon and their effect might increase.

Today, patent aggregating companies that rely on patent enforcement are mainly active in the US. Due to several system factors, the US system is a better basis for this type of business model. In the European Union, a pan-European patent court has been discussed for several years now. The establishment of such a pan-European court could change the legal situation in Europe. Therefore, it could foster the way of increasing patent enforcement and aggregating companies that operate based on patent infringements.

In addition to innovation transfer and patent enforcement, patent aggregating companies that focus on a non-commercial use of patents will strengthen. In the pharmaceutical industry or in green technology, this type of patent aggregating company will become important. In times of natural catastrophes, such as floods and droughts that destroy the harvest of whole countries, tools to repair damages, prevent following losses, control resulting diseases, and relieve the distress become increasingly important. As R&D expenditures for drugs are high and the prices in developing countries must be low to be affordable for low incomes, consortia for the development of drugs distributed in developing countries seems a good alternative to handle high R&D costs. In addition, green technology has to be developed based on a global perspective to help all affected regions.

Patent aggregating companies have emerged during the last decade and are still changing, developing, and vanishing. As the market for patents and technologies has grown to major importance, patent aggregating companies will stay major players in this market. Depending on the development of the market regarding efficiency, transparency, and the legal system, several more business models will emerge that use these different characteristics. Producing companies have to accept the existence of patent aggregating companies in their diversity and learn how to utilize them optimally.

References

Acacia Research Corporation. (2010). *Form 10-K Annual Report: Fiscal year ended December 31, 2010.*

Adam, Y., Ong, C. H., & Pearson, A. W. (1988). Licensing as an Alternative to Foreign Direct Investment. *Journal of Product Innovation Management, 5*(1), 32–49.

AIPLA. (2011). *Report of the Economic Survey 2011.* Retrieved from http://www.aipla.org/learningcenter/library/books/econsurvey/2011/Documents/AIPLA%202011%20Report-%20JULY%202011_final.pdf

Akerlof, G. A. (1970). The market for "lemons": Quality, uncertainty and the market mechanism. *Quarterly Journal of Economics, 84*(3), 488–500.

Allied Security Trust. (2010). *Allied Security Trust Announces Availability of Major Patent Portfolio.* Retrieved from http://www.bloomberg.com/apps/news?pid=conewsstory&tkr=NIPNY:US&sid=a_.V6TdpLql0

Anand, B. N., & Khanna, T. (2000). The structure of licensing contracts. *Journal of Industrial Economics, 48*(1), 103–135.

Andersen-Gott, M., Ghinea, G., & Bygstad, B. (in press). Why do commercial companies contribute to open source software? *International Journal of Information Management.*

Anderson, B. (1979). Acquiring and selling technology - marketing techniques. *Research Management, 22*(3), 26–28.

Aoki, R., & Hu, J.-L. (2003). Time Factors of Patent Litigation and Licensing. *Journal of Institutional and Theoretical Economics JITE, 159*(2), 280–301.

Aoki, R., & Schiff, A. (2008). Promoting access to intellectual property: patent pools, copyright collectives, and clearinghouses. *R&D Management, 38*(2), 189–204.

Aronoff, R. (2011). The state of the US IP Marketplace 2010-2011. *The Computer & Internet Lawyer, 28*(2), 1–5.

Arora, A. (1995). Licensing Tacit Knowledge: Intellectual Property Rights And The Market For Know-How. *Economics of Innovation and New Technology, 4*(1), 41–60.

Arora, A., & Gambardella, A. (2010a). Ideas for rent: an overview of markets for technology. *Industrial & Corporate Change, 19*(3), 775–803.

Arora, A., & Gambardella, A. (2010b). The Market for Technology. In B. Hall & N. Rosenberg (Eds.), *Handbook of the Economics of Innovation.* North Holland: Elsevier Science & Technology.

Arora, A., Fosfuri, A., & Gambardella, A. (2001a). *Markets for technology: The economics of innovation and corporate strategy.* Cambridge, MA: MIT Press.

Arora, A., Fosfuri, A., & Gambardella, A. (2001b). Markets for Technology and their Implications for Corporate Strategy. *Industrial & Corporate Change, 10*(2), 419–451.

Arundel, A. & Patel, P. (2003). *Strategic patenting* (Background report for the Trend Chart Policy Benchmarking). Luxembourg.

Arundel, A., van de Paal, G., & Soete, L. (1995). *Innovation Strategies of Europe's Largest Industrial Firms* (PACE Report).

Athreye, S., & Cantwell, J. (2007). Creating competition? Globalisation and the emergence of new technology producers. *Research Policy, 36*(2), 209–226.

Atuahene-Gima, K. (1992). Inward technology licensing as an alternative to internal R&D in new product development: A conceptual framework. *Journal of Product Innovation Management, 9*(2), 156–167.

Avancept. (2001). *A Study Of The Intellectual Ventures Portfolio In the United States: Patents & Applications* (2nd Edition, Version 2.4, Sample Report).

Bader, M. A. (2006). *Intellectual property management in R&D collaborations: The case of the service industry sector*. Heidelberg: Physica-Verlag.

Bader, M. A., Gassmann, O., Ziegler, N., & Ruether, F. (in press). Getting the most out of your IP – patent management along its life cycle. *Drug Discovery Today*.

Ball, G. & Kesan, J. P. (2009). *Transaction Costs and Trolls: Strategic Behavior by Individual Inventors, Small Firms and Entrepreneurs in Patent Litigation* (University of Illinois Law & Economics Research Paper No. LE09-005; Illinois Public Law Research Paper No. 08-21). Retrieved from http://ssrn.com/abstract=1337166

Barney, J. (1991). Firm Resources and Sustained Competitive Advantage. *Journal of Management, 17*(1), 99–120.

Barry, C., Johnston, A., Arad, R., Stainback, D., Ansell, L., & Arnold, M. (2010). *The continued evolution of patent damages law: 2010 patent litigation study* (Applied social research methods series). Retrieved from http://www.pwc.com/us/en/forensic-services/publications/2010-patent-litigation-study.jhtml

Becker, B. (2003). *Corporate incubators: Potentials, typology and principles*. Sankt Gallen: Dissertation.

Benassi, M., & Di Minin, A. (2009). Playing in between: patent brokers in markets for technology. *R&D Management, 39*(1), 68–86.

Benassi, M., Corsaro, D., & Geenen, G. (2010). *Are patent brokers a possible first best?* (Working Paper no. 2010-11).

Berneman, L., Cockburn, I., Agrawal, A., & Iyer, S. (2009). U.S./Canadian Licensing In 2007-08: Survey Results. *Les Nouvelles*, (March), 1–8.

Bessant, J., & Rush, H. (1995). Building bridges for innovation: the role of consultants in technology transfer. *Research Policy, 24*(1), 97–114.

Bessen, J., Ford, J., & Meurer, M. (2011). *The private and social costs of patent trolls* (Boston University School of Law Working Paper No. 11-45). Retrieved from http://www.bu.edu/law/faculty/scholarship/workingpapers/2011.html

Bessler, W., Bittelmeyer, C., & Lipfert, S. (2003). Zur Bedeutung von wissensbasierten immateriellen Vermögensgegenständen für die Bewertung und Finanzierung von kleinen und mittleren Unternehmen: Ein Überblick. In J.-A. Meyer (Ed.), *Unternehmensbewertung und Basel II in kleinen und mittleren Unternehmen* (1st ed., pp. 309–334). Lohmar: Eul.

Bianchi, M., Chiaroni, D., Chiesa, V., & Frattini, F. (2011a). Exploring the role of human resources in technology out-licensing:an empirical analysis of biotech newtechnology-based firms. *Technology Analysis & Strategic Management*, *23*(8), 825–849.

Bianchi, M., Chiaroni, D., Chiesa, V., & Frattini, F. (2011b). Organizing for external technology commercialization: evidence from a multiple case study in the pharmaceutical industry. *R&D Management*, *41*(2), 120–137.

Birkenmeier, B. U. (2003). *Externe Technologie-Verwertung: Eine komplexe Aufgabe des Integrierten Technologie-Managements.* Zürich: Diss., Technische Wissenschaften ETH Zürich, Nr. 15140, 2003.

Blind, K., & Thumm, N. (2004). Interrelation between patenting and standardisation strategies: empirical evidence and policy implications. *Research Policy*, *33*(10), 1583–1598.

Blind, K., Cremers, K., & Mueller, E. (2009). The influence of strategic patenting on companies' patent portfolios. *Research Policy*, *38*(2), 428–436.

Blind, K., Edler, J., Frietsch, R., & Schmoch, U. (2006a). Motives to patent: Empirical evidence from Germany. *Research Policy*, *35*(5), 655–672.

Blind, K., Edler, J., Frietsch, R., & Schmoch, U. (2006b). *Scope and nature of the patent surge: A view from Germany.*

Bloomberg News. (2007). *Microsoft Settles a Dispute Over a Feature in Its Browse.* Retrieved from http://www.nytimes.com/2007/08/31/technology/31soft.html

Boyens, K. (1998). *Externe Verwertung von technologischem Wissen.* Wiesbaden: Deutscher Universitäts-Verlag.

Bryant, T. A., & Reenstra-Bryant, R. A. (1998). Technology brokers in the North American software industry: Getting the most out of mismatched. *International Journal of Technology Management*, *16*(1-3), 281.

Burt, R. S. (1995). *Structural holes: The social structure of competition.* Cambridge, MA: Harvard University Press.

Burt, R. S. (2005). *Brokerage and closure: An introduction to social capital.* Oxford: Oxford University Press.

Business Wire. (2011). *Patentportfolio 2 S.à r.l. Completes Purchase of 400 Patent Assets from BT Group PLC and Retains IP Navigation Group (Europe) for Patent Monetization Program.* Retrieved from http://www.businesswire.com/news/home/20110803006954/en/Patentportfolio-2-S.%C3%A0-r.l.-Completes-Purchase-400

Callon, M., & Muniesa, F. (2005). Economic Markets as Calculative Collective Devices. *Organization Studies, 26*(8), 1229–1250.

Carlsson, B., Dumitriu, M., Glass, J., Nard, C., & Barrett, R. (2008). Intellectual property (IP) management: organizational processes and structures, and the role of IP donations. *The Journal of Technology Transfer, 33*(6), 549–559. doi:10.1007/s10961-008-9082-2

Caves, R. E., Crookell, H., & Killing, J. P. (1983). The imperfect market for technology licenses. *Oxford Bulletin of Economics & Statistics, 45*(3), 249–267.

Caves, R. E., Whinston, M. D., & Hurwitz, M. A. (1991). Patent expiration, entry, and competition in the U.S. pharmaceutical industry. *Brookings Papers on Economic Activity*, 1–48.

Cesaroni, F. (2004). Technological outsourcing and product diversification: do markets for technology affect firms' strategies? *Research Policy, 33*(10), 1547–1564.

Cesaroni, F., & Mariani, M. (2001). The market for knowledge in the chemical sector. In B. Guilhon (Ed.), *Technology and Markets for Knowledge. Knowledge Creation, Diffusion and Exchange within a Growing Economy* (pp. 71–97). Dordrecht: Kluwer Academic Publishers.

Chesbrough, H. (2003a). The Logic of Open Innovation: Managing Intellectual Property. *California Management Review, 45*(3), 33–58.

Chesbrough, H. W. (2003b). The Era of Open Innovation. *MIT Sloan Management Review, 44*(3), 35–41.

Chesbrough, H. W. (2003c). The governance and performance of Xerox's technology spin-off companies. *Research Policy, 32*(3), 403–421.

Chesbrough, H. W. (2006). *Open business models: How to thrive in the new innovation landscape*. Boston, MA: Harvard Business School Press.

Chien, C. V. (2009). Of Trolls, Davids, Goliaths, and Kings: Narratives and Evidence in the Litigation of high-tech Patents. *North Carolina Law Review, 87*, 1571–1615.

Christensen, J. F., Olesen, M. H., & Kjær, J. S. (2005). The industrial dynamics of Open Innovation: Evidence from the transformation of consumer electronics. *Research Policy, 34*(10), 1533–1549.

Cockburn, I. M., MacGarvie, M. J., & Müller, E. (2010). Patent thickets, licensing and innovative performance. *Industrial & Corporate Change, 19*(3), 899–925.

Cohen, W. M., & Levinthal, D. A. (1990). Absorptive Capacity: A New Perspective on Learning and Innovation. *Administrative Science Quarterly, 35*(1), 128–152.

Cohen, W. M., Goto, A., Nagata, A., Nelson, R. R., & Walsh, J. P. (2002). R&D spillovers, patents and the incentives to innovate in Japan and the United States. *Research Policy, 31*(8-9), 1349–1367.

Cohen, W. M., Nelson, R. R., & Walsh, J. P. (2000). *Protecting Their Intellectual Assets: Appropriability Conditions and Why U.S. Manufacturing Firms Patent (or*

Not) (NBER Working Paper No. 7552). Retrieved from http://www.nber.org/papers/w7552

Conner, K. R. (1995). Obtaining Strategic Advantage from Being Imitated: When Can Encouraging "Clones" Pay? *Management Science, 41*(2), 209–225.

Contractor, F. J. (1980). The "profitability" of technology licensing by U. S. multinationals: A framework for analysis and an empirical study. *Journal of International Business Studies, 11*(2), 40–63.

Cusumano, M. A. (2004). Reflections on Free and Open Software. *Communications of the ACM, 47*(10), 25–27.

Dahlander, L. (2005). Appropriation and Appropriability in open source software. *International Journal of Innovation Management, 9*(3), 259–285.

Dahlander, L., & Wallin, M. W. (2006). A man on the inside: Unlocking communities as complementary assets. *Research Policy, 35*(8), 1243–1259.

Davenport, S., Carr, A., & Bibby, D. (2002). Leveraging talent: spin–off strategy at Industrial Research. *R&D Management, 32*(3), 241–254. doi:10.1111/1467-9310.00257

Davis, J. L., & Harrison, S. S. (2001). *Edison in the boardroom: How leading companies realize value from their intellectual assets*. Hoboken, NJ: John Wiley & Sons.

Davis, L. (2004). Intellectual property rights, strategy and policy. *Economics of Innovation & New Technology, 13*(5), 399–415.

Davis, R. (2008). Failed Attempts to Dwarf the Patent Trolls: Permanent Injunctions in Patent Infringement Cases Under The Proposed Patent Reform Act of 2005 and Ebay v. Mercexchange. *Cornell Journal of Law and Public Policy, 17*(2), 431–452.

Dhanaraj, T. (in press). A limited revolution: The distributional consequences of Open Source Software in North America. *Technological Forecasting and Social Change*.

D'Iribarne, P. (1996). The Usefulness of an Ethnographic Approach to the International Comparison of Organizations. *International Studies of Management & Organization, 26*(4), 30–47.

Duguet, E., & Kabla, I. (1998). Appropriation strategy and the motivations to use the patent system: an econometric analysis at the firm level in French manufacturing. *Annales d'Economie et de Statistique*, (49/50), 289–327.

Ehrhardt, M. (2004). Network effects, standardisation and competitive strategy: how companies influence the emergence of dominant designs. *International Journal of Technology Management, 27*(2/3), 272–294.

Eisenhardt, K. (1989a). Building Theories from Case Study Research. *Academy of Management Review, 14*(4), 532–550.

Eisenhardt, K. M. (1989b). Agency Theory: An Assessment and Review. *Academy of Management Review, 14*(1), 57–74.

Elton, J. J., Shah, B. R., & Voyzey, J. N. (2002). Intellectual property: Partnering for profit. *McKinsey Quarterly*, (4), 59–67.

Ensthaler, J., & Strübbe, K. (2006). *Patentbewertung: Ein Praxisleitfaden zum Patentmanagement* (1st ed.). Berlin: Springer.

EPO. (2009). *Patent information.* Retrieved from http://documents.epo.org/projects/babylon/eponet.nsf/0/43c7e3178f759014c125728b0044e305/$FILE/patent_information_en.pdf

EPO, OECD, & UKIPO. (2006). *Patents: Realising and securing value: Executive summary.* Conference organised by the EPO, the OECD and the UK Patent Office, London, 21 November 2006.

Ernst, H. (1995). Patenting strategies in the German mechanical engineering industry and their relationship to company performance. *Technovation, 15*(4), 225–240.

Ernst, H. (2001). Patent applications and subsequent changes of performance: evidence from time-series cross-section analyses on the firm level. *Research Policy, 30*(1), 143–157.

Ewing, T. (2010). Inside the world of public auctions. *Intellectual Asset Management Magazine, 42*, 63–70.

Fiol, C. M., & Lyles, M. A. (1985). Organizational Learning. *Academy of Management Review, 10*(4), 803–813.

Fischer, T. & Henkel, J. (2009). *Patent Trolls on Markets for Technology: An Empirical Analysis of Trolls Patent Acquisitions* (Working paper). Retrieved from http://papers.ssrn.com/sol3/papers.cfm?abstract_id=1523102

Fishman, E. (2003). Securitization of IP Royalty Streams: Assessing the Landscape. *Technology Access Report*, (September), 6–7.

Fitzgerald, B. (2006). The Transformation of Open Source Software. *MIS Quarterly, 30*(3), 587–598.

Ford, D. (1985). The management and marketing of technology. In R. Lamb & P. Shrivastava (Eds.), *Advances in strategic management. A research annual* (pp. 103–134). London: Jai Press.

Ford, D., & Ryan, C. (1977). The marketing of technology. *European Journal of Marketing, 11*(6), 369–383.

Ford, D., & Ryan, C. (1981). Taking technology to market. *Harvard Business Review, 59*(2), 117–126.

Gambardella, A. (2002). Successes and Failures in the Markets for Technology. *Oxford Review of Economic Policy, 18*(1), 52–62.

Gambardella, A., Giuri, P., & Luzzi, A. (2007). The market for patents in Europe. *Research Policy, 36*(8), 1163–1183.

Gambardella, A., Harhoff, D., & Verspagen, B. (2008). The value of European patents. *European Management Review, 5*(2), 69–84.

Gans, J. S., & Stern, S. (2003). The product market and the market for "ideas": commercialization strategies for technology entrepreneurs. *Research Policy, 32*(2), 333–350.

Gans, J. S., Hsu, D. H., & Stern, S. (2008). The Impact of Uncertain Intellectual Property Rights on the Market for Ideas: Evidence from Patent Grant Delays. *Management Science, 54*(5), 982–997.

Gans, J., & Stern, S. (2010). Is there a market for ideas? *Industrial and Corporate Change, 19*(3), 805–837.

Gassmann, O. (2006). Opening up the innovation process: towards an agenda. *R&D Management, 36*(3), 223–228.

Gassmann, O., & Bader, M. A. (2011). *Patentmanagement: Innovationen erfolgreich nutzen und schützen* (3rd ed.). Berlin: Springer.

Gassmann, O., Kobe, C., & Voit, E. (2001). *High-Risk-Projekte: Quantensprünge in der Entwicklung erfolgreich managen*. Berlin: Springer.

Geradin, D., Layne-Farrer, A., & Padilla, A. J. (2011). *Elves or trolls? The role of non-practicing patent owners in the innovation economy* (TILEC Discussion Paper No. 2008-018). Retrieved from http://ssrn.com/abstract=1136086

Gilbert, R. J., & Newbery, D. M. G. (1982). Preemptive Patenting and the Persistence of Monopoly. *American Economic Review, 72*(3), 514–526.

Giummo, J. (2010). German employee inventors' compensation records: A window into the returns to patented inventions. *Research Policy, 39*(7), 969–984.

Giuri, P., Mariani, M., Brusoni, S., Crespi, G., Francoz, D., Gambardella, A., … (2007). Inventors and invention processes in Europe: Results from the PatVal-EU survey. *Research Policy, 36*(8), 1107–1127.

Golden, J. M. (2007). "Patent Trolls" and Patent Remedies. *Texas Law Review*, pp. 2111–2161.

Graevenitz, G. von, Wagner, S., & Harhoff, D. (2008). *Incidence and growth of patent thickets: the impact of technological opportunities and complexity* (Working paper). Retrieved from http://elsa.berkeley.edu/~bhhall/e222spring07_files/HarhoffWagnervonGraevenitz08_patent_thickets.pdf

Granstrand, O. (2000). *The economics and management of intellectual property: Towards intellectual capitalism*. Northamptom: Edward Elgar Publishing.

Granstrand, O. (2004). The economics and management of technology trade: Towards a pro-licensing era? *International Journal of Technology Management, 27*(2/3), 209–240.

Grant, R. M. (1996). Toward a Knowledge-Based Theory of the Firm. *Strategic Management Journal, 17*(Winter Special Issue), 109–122.

Gredel, D., Kramer, M., & Bend, B. (in press). Patent-based investment funds as innovation intermediaries for SMEs: In-depth analysis of reciprocal interactions, motives and fallacies. *Technovation*.

Gregory, J. K. (2007). The Troll next Door. *The John Marshall Review of Intellectual Property Law, 6*(2), 292–309.

Griliches, Z. (1990). Patent Statistics as Economic Indicators: A Survey. *Journal of Economic Literature, 28*(4), 1661–1707.

Grindley, P. C., & Teece, D. J. (1997). Managing Intellectual Capital: Licensing and Cross-Licensing in Semiconductors and Electronics. *California Management Review, 39*(2), 8–41.

Gruenwedel, E. (2011). *Analyst: Blu-ray to Reach 74% Market Share by 2017*. Home Media Magazine. Retrieved from http://www.homemediamagazine.com/blu-ray-disc/analyst-blu-ray-reach-74-market-share-2017-14496

Guilhon, B. (2001). The Emergence of the Quasimarkets for Knowledge. In B. Guilhon (Ed.), *Technology and Markets for Knowledge. Knowledge Creation, Diffusion and Exchange within a Growing Economy* (pp. 21–40). Dordrecht: Kluwer Academic Publishers.

Guilhon, B. (2004). Markets for knowledge: problems, scope, and economic implications. *Economics of Innovation and New Technology, 13*(2), 165–181.

Hägerstrand, T. (1952). *The propagation of innovation waves. Lund studies in geography / B: Vol. 4*. Lund: Department of Geography.

Hall, B. (2005). Exploring the Patent Explosion. In A. N. Link & F. M. Scherer (Eds.), *Essays in Honor of Edwin Mansfield* (pp. 195–208). New York, NY: Springer.

Hall, B. H., & Ziedonis, R. H. (2001). The patent paradox revisited: an empirical study of patenting in the U.S. semiconductor industry, 1979-1995. *RAND Journal of Economics, 32*(1), 101–128.

Hall, R. (1992). The strategic analysis of intangible resources. *Strategic Management Journal, 13*(2), 135–144.

Hanel, P. (2006). Intellectual property rights business management practices: A survey of the literature. *Technovation, 26*(8), 895–931.

Hargadon, A., & Sutton, R. I. (1997). Technology Brokering and Innovation in a Product Development Firm. *Administrative Science Quarterly, 42*(4), 716–749.

Harhoff, D., & Reitzig, M. (2004). Determinants of opposition against EPO patent grants: the case of biotechnology and pharmaceuticals. *International Journal of Industrial Organization, 22*(4), 443–480. doi:10.1016/j.ijindorg.2004.01.001

Harhoff, D., Scherer, F. M., & Vopel, K. (2003a). Citations, family size, opposition and the value of patent rights. *Research Policy, 32*(8), 1343–1363.

Harhoff, D., Scherer, F. M., & Vopel, K. (2003b). Exploring the tail of patented invention value distribution. In O. Granstrand (Ed.), *Economics, Law and Intellectual Property. Seeking Strategies for Research and Teaching in a Developing Field* (pp. 279–308). Dordrecht: Kluwer Academic Publishers.

Harris, R. G. (2001). The knowledge-based economy: Intellectual origins and new economic perspectives. *International Journal of Management Reviews, 3*(1), 21–40.

Henkel, J. & Reitzig, M. (2007). *Patent Sharks and the Sustainability of Value Destruction Strategies* (Working paper). Retrieved from http://papers.ssrn.com/sol3/papers.cfm?abstract_id=985602

Henkel, J., & Reitzig, M. (2008). Patent Sharks. *Harvard Business Review, 86*(6), 129–133.

Hetzel, D. (2010). Embracing the new IP reality. *Intellectual Asset Management Magazine, 41*, 29–34.

Hippel, E. von. (1988). *The sources of innovation.* New York, NY: Oxford University Press.

Hoffmann, W. H., & Schlosser, R. (2001). Success Factors of Strategic Alliances in Small and Medium-sized Enterprises-An Empirical Survey. *Long Range Planning, 34*(3), 357–381.

Holden, P. (2011). New models in response to changes in the global IP market. *Intellectual Asset Management Magazine, 48*, 37–42.

Howells, J. (2006). Intermediation and the role of intermediaries in innovation. *Research Policy, 35*(5), 715–728.

Huber, G. P. (1991). Organizational learning: The contributing process and the literatures. *Organization Science, 2*(1), 88–115.

Humphrey, J., West Jr., K., & Sommer, A. (1992). Vitamin A deficiency and attributable mortality among under-5-year-olds. *Bulletin of the World Health Organization, 70*(2), 225–232.

Intellectual Property Today. (2010). *VisEn Acquires Key Fluorescence Agent Intellectual Property Portfolio and Technology Platforms From Bayer Schering Pharma.* Retrieved from http://www.iptoday.com/news-archived-article.asp?id=5075&type=business

Intellectual Ventures. (2009). *Intellectual Ventures Acquires Transmeta Patent Portfolio.* Retrieved from http://www.intellectualventures.com/newsroom/pressreleases/archive/09-01-28/Intellectual_Ventures_Acquires_Transmeta_Patent_Portfolio.aspx

Janicke, P. M., & Ren, L. (2006). Who Wins Patent Infringement Cases? *American Intellectual Property Law Association Quarterly Journal, 34*(1), 1–43.

Jarboe, K. & Furrow, R. (2008). *Intangible Asset Monetization: The Promise and the Realty* (Athena Alliance Working Paper No. 3). Retrieved from http://athenaalliance.org/pdf/IntangibleAssetMonetization.pdf

Jensen, M. C., & Meckling, W. H. (1976). Theory of the firm: Managerial behavior, agency costs and ownership structure. *Journal of Financial Economics, 3*(4), 305–360.

Johnson, J., Leonard, G., Meyer, C., & Serwin, K. (2007). Don't feed the trolls? *Les Nouvelles, 52*(3), 487–495.

Jones, G. K., Lanctot, A., & Teegen, H. J. (2001). Determinants and performance impacts of external technology acquisition. *Journal of Business Venturing, 16*(3), 255–283.

Jorion, P. (2009). *Financial risk manager handbook* (5th ed.). Hoboken, NJ: John Wiley & Sons.

Kale, P., Dyer, J. H., & Singh, H. (2002). Alliance Capability, Stock Market Response, and Long Term Alliance Success: The Role of the Alliance Function. *Strategic Management Journal, 23*(8), 747.

Kamiyama, S., Sheehan, J., & Martinez, C. (2006). *Valuation and Exploitation of Intellectual Property* (OECD Science, Technology and Industry Working Papers 2006/5). Retrieved from http://www.oecd.org/dataoecd/62/52/37031481.pdf

Kash, D., & Kingston, W. (2001). Patents in a world of complex technologies. *Science and Public Policy, 28*(1), 11–22.

Kelley, A. (2011). Practicing in the Patent Marketplace. *The University of Chicago Law Review, 78*(1), 115–137.

Keupp, M. M., & Gassmann, O. (2009). Determinants and archetype users of open innovation. *R&D Management, 39*(4), 331–341.

Klerkx, L., & Leeuwis, C. (2009). Establishment and embedding of innovation brokers at different innovation system levels: Insights from the Dutch agricultural sector. *Technological Forecasting and Social Change, 76*(6), 849–860.

Kline, D. (2003). Sharing the Corporate Crown Jewels. *MIT Sloan Management Review, 44*(3), 89–93.

Kogut, B., & Zander, U. (1992). Knowledge of the firm, combinative capabilities, and the replication of technology. *Organization Science, 3*(3), 383–397.

Kogut, B., & Zander, U. (1993). Knowledge of the Firm and the Evolutionary Theory of the Multinational Corporation. *Journal of International Business Studies, 24*(4), 625–645.

Koruna, S. M. (2004). External technology commercialisation policy guidelines. *International Journal of Technology Management, 27*(2/3), 241–254.

Krattiger, A., & Potrykus, I. (2007). Golden Rice: A Product-Development Partnership in Agricultural Biotechnology and Humanitarian Licensing. In A. Krattiger (Ed.), *Executive Guide to Intellectual Property Management in Health and Agricultural Innovation: A Handbook of Best Practices* (pp. 11–14). Oxford: MIHR.

Krogh, G. von, & Hippel, E. von. (2003). Special issue on open source software development: Open Source Software Development. *Research Policy, 32*(7), 1149–1157.

Kromrey, H. (1998). *Empirische Sozialforschung: Modelle und Methoden der Datenerhebung und Datenauswertung* (8th ed.). Opladen: Leske + Budrich.

Lamnek, S. (1995). *Methoden und Techniken* (3rd ed.). München: Psychologie-Verlag-Union.

Lamoreaux, N., & Sokoloff, K. (2007). The Market for Technology and the Organization of Invention in U.S. History. In E. Sheshinski, R. J. Strom, & W. J. Baumol (Eds.), *Entrepreneurship, innovation, and the growth mechanism of the free-enterprise economies* (pp. 213–243). Princeton, NJ: Princeton Univ. Press.

Lane, P. J., Koka, B. R., & Pathak, S. (2006). The Reification of Absorptive Capacity: A Critical Review and Rejuvenation of the Construct. *Academy of Management Review, 31*(4), 833–863.

Lanjouw, J. O., & Schankerman, M. (2001). Characteristics of patent litigation: A window on competition. *The Rand Journal of Economics, 32*(1), 129–151.

Lanjouw, J. O., & Schankerman, M. (2004). Patent Quality and Research Productivity: Measuring Innovation with Multiple Indicators. *The Economic Journal, 114*(495), 441–465. doi:10.1111/j.1468-0297.2004.00216.x

Laurie, R. (2006). *Integrating an Intellectual Property Strategy into Your Business Plan* (Newcom 2006 Winter School). Torino. Retrieved from http://ip-strategy.com/downloads/NEWCOM_2006_Integrating_IP_Strategy_into_Bus_Plan.pdf

Laurie, R. (2007). *Best practice for buying, selling and licensing patents* (IP Law Seminars International). Palo Alto. Retrieved from http://www.ip-strategy.com/downloads/LSI_Strategic_Patent_Acquisition.pdf

Lavie, D. (2006). The Competitive Advantage of Interconnected Firms: An Extension of the Resource-Based View. *Academy of Management Review, 31*(3), 638–658.

Layne-Farrar, A. & Schmidt, K. (2009). *Licensing complementary patents: 'Patent trolls', market structure, and 'excessive' royalties* (Working paper). Retrieved from http://works.bepress.com/anne_layne_farrar/6/

Leiponen, A., & Byma, J. (2009). If you cannot block, you better run: Small firms, cooperative innovation, and appropriation strategies. *Research Policy, 38*(9), 1478–1488.

Lemley, M. A. (2007). *Are Universities Patent Trolls?* (Stanford Public Law Working Paper No. 980776). Retrieved from http://ssrn.com/abstract=980776

Levin, R. C. (1986). A New Look at the Patent System. *American Economic Review, 76*(2), 199–202.

Levin, R. C., Klevorick, A. K., Nelson, R. R., & Winter, S. G. (1987). Appropriating the Returns from Industrial Research and Development. *Brookings Papers on Economic Activity, Special Issue*(3), 783–820.

Levitt, B., & March, J. G. (1988). Organizational Learning. *Annual Review of Sociology, 14*(1), 319–338.

Levitt, T. (1965). Exploit the Product Life Cycle. *Harvard Business Review, 43*(6), 81–94.

Lichtenthaler, E. (2004). Organising the external technology exploitation process: current practices and future challenges. *International Journal of Technology Management, 27*(2), 255–271.

Lichtenthaler, U. (2005). External commercialization of knowledge: Review and research agenda. *International Journal of Management Reviews, 7*(4), 231–255.

Lichtenthaler, U. (2006). Technology exploitation strategies in the context of open innovation. *International Journal of Intelligence and Planning, 2*(1), 1–21.

Lichtenthaler, U. (2007a). Managing external technology commercialisation: A process perspective. *International Journal of Technology Marketing, 2*(3), 225–242.

Lichtenthaler, U. (2007b). The Drivers of Technology Licensing: An Industry Comparison. *California Management Review, 49*(4), 67–89.

Lichtenthaler, U. (2007c). Trading intellectual property in the new economy. *International Journal of Intellectual Property Management, 1*(3), 241–252.

Lichtenthaler, U. (2008a). External technology commercialisation projects: Objectives, processes and a typology. *Technology Analysis & Strategic Management, 20*(4), 483–501.

Lichtenthaler, U. (2008b). Leveraging technology assets in the presence of markets for knowledge. *European Management Journal, 26*(2), 122–134.

Lichtenthaler, U. (2010). Determinants of proactive and reactive technology licensing: A contingency perspective. *Research Policy, 39*(1), 55–66.

Lichtenthaler, U. (2011). The evolution of technology licensing management: identifying five strategic approaches. *R&D Management, 41*(2), 173–189.

Lichtenthaler, U., & Ernst, H. (2006). Attitudes to externally organising knowledge management tasks: a review, reconsideration and extension of the NIH syndrome. *R&D Management, 36*(4), 367–386.

Lichtenthaler, U., & Ernst, H. (2007). Developing reputation to overcome the imperfections in the markets for knowledge. *Research Policy, 36*(1), 37–55.

Lichtenthaler, U., & Ernst, H. (2008a). Innovation Intermediaries: Why Internet Marketplaces for Technology Have Not Yet Met the Expectations. *Creativity & Innovation Management, 17*(1), 14–25.

Lichtenthaler, U., & Ernst, H. (2008b). Intermediary Services in the Markets for Technology: Organizational Antecedents and Performance Consequences. *Organization Studies, 29*(7), 1003–1035.

Lichtenthaler, U., & Ernst, H. (2009). The Role of Champions in the External Commercialization of Knowledge. *Journal of Product Innovation Management, 26*(4), 371–387.

Lichtenthaler, U., & Lichtenthaler, E. (2009). A Capability-Based Framework for Open Innovation: Complementing Absorptive Capacity. *Journal of Management Studies, 46*(8), 1315–1338.

Lipfert, S., & Ostler, J. (2008). Fonds und Auktionen: Neue Formen der Patentverwertung. In T. Tiefel (Ed.), *Gewerbliche Schutzrechte im Innovationsprozess* (pp. 85–106). Wiesbaden: Deutscher Universitäts-Verlag.

Lipfert, S., & Scheffer, G. von. (2006). Europe's first patent value fund. *Intellectual Asset Management Magazine, 15*, 15–18.

Lippert, T. (2011). *Alpha Patentfonds 1 - Alpha Patentfonds 2 - Wie geht es weiter?* (Aktionsbund Aktiver Anlegerschutz). Retrieved from http://www.aktionsbund.de/app/themes/news/alpha-patentfonds-1-alpha-patentfonds-2-wie-geht-es-weiter-658

Lopez-Vega, H. (2009). *How demand-driven Technological Systems of Innovation work? The role of Intermediary organizations* (Conference Paper). Retrieved from http://www2.druid.dk/conferences/viewpaper.php?id=4244&cf=33

Luman III, J., & Dodson, C. (2006). No longer a Myth, the Emergence of the Patent Troll: Stifling Innovation, Increasing Litigation, and Extorting Billions. *Intellectual Property & Technology Law Journal, 18*(5), 12–16.

Lynn, L. H., Mohan Reddy, N., & Aram, J. D. (1996). Linking technology and institutions: the innovation community framework. *Research Policy, 25*(1), 91–106.

Magliocca, G. N. (2007). Blackberries and Barnyards: Patent Trolls and the Perils of Innovation. *Notre Dame Law Review, 82*(5), 1809–1838.

Malackowski, J. E., Cardoza, K., Gray, C., & Conroy, R. (2007). The Intellectual Property Marketplace: Emerging Transaction and Investment Vehicles. *The Licensing Journal, 27*(2), 1–11.

Mansfield, E. (1986). Patents and Innovation: An Emprical Study. *Management Science, 32*(2), 173–181.

Marcy, W. (1979). Acquiring and Selling Technology-Licensing Do's and Don'ts. *Research Management, 22*(3), 18–21.

Mathews, J. A. (2003). Strategizing by firms in the presence of markets for resources. *Industrial & Corporate Change, 12*(6), 1157–1193.

Mazzoleni, R., & Nelson, R. R. (1998). The benefits and costs of strong patent protection: a contribution to the current debate. *Research Policy, 27*(3), 273–284.

McCurdy, D., & Reohr, C. (2008). A new tool for a new kind of patent adversary. *Intellectual Asset Management Magazine, 32*, 31–35.

McDonough III, J. F. (2006). The Myth of the Patent Troll: An Alternative View of the Function of Patent Dealers in an Idea. *Emory Law Journal, 56*(1), 189–228.

Merges, R. P. (2009). The trouble with trolls: Innovation, rent-seeking, and patent law reform. *Berkeley Technology Law Journal, 24*(4), 1583–1614.

Merges, R. P., & Nelson, R. R. (1990). On the Complex Economics of Patent Scope. *Columbia Law Review, 90*(4), 839–916.

Merriam, S. B. (1998). *Qualitative research and case study applications in education* (2nd ed.). San Francisco, CA: Jossey-Bass Publications.

Miles, M. B. (1979). Qualitative Data as an Attractive Nuisance: The Problem of Analysis. *Administrative Science Quarterly, 24*(4), 590–601.

Miles, M. B., & Huberman, M. A. (2004). *Qualitative data analysis: An expanded sourcebook* (2nd ed.). Thousand Oaks, CA: SAGE Publications.

Millien, R., & Laurie, R. (2008). Meet the middleman. *Intellectual Asset Management Magazine, 28*, 53–58.

Mintzberg, H. (2007). Developing theory about the development of theory. In K. G. Smith & M. A. Hitt (Eds.), *Great Minds in Management: The Process of Theory Development. The Process of Theory Development*. Oxford: Oxford University Press.

Mohr, J., & Spekman, R. (1994). Characteristics of partnership success: Partnership attributes, communication behavior, and conflict resolution techniques. *Strategic Management Journal, 15*(2), 135–152.

Monk, A. H. B. (2009). The emerging market for intellectual property: drivers, restrainers, and implications. *Journal of Economic Geography, 9*(4), 469–491.

Morgan, E., & Crawford, N. (1996). Technology broking activities in Europe - a survey. *International Journal of Technology Management*, *12*(3), 360–367.

Murphy, D. (2008). *Understanding risk: The theory and practice of financial risk management*. Boca Raton, FL: Chapman & Hall.

Nambisan, S., & Sawhney, M. (2007). A Buyer's Guide to the Innovation Bazaar. *Harvard Business Review*, *85*(6), 109–118.

Narin, F., Noma, E., & Perry, R. (1987). Patents as indicators of corporate technological strength. *Research Policy*, *16*(2-4), 143–155.

Noel, M. & Schankerman, M. (2011). *Strategic Patenting and Software Innovation* (LSE STICERD Research Paper No. EI43). Retrieved from http://papers.ssrn.com/sol3/papers.cfm?abstract_id=1158320

Nonaka, I. (1994). A Dynamic Theory of Organizational Knowledge Creation. *Organization Science*, *5*(1), 14–37.

Nonaka, I., Toyama, R., & Nagata, A. (2000). A firm as a knowledge-creating entity: a new perspective on the theory of the firm. *Industrial & Corporate Change*, *9*(1), 14–37.

OECD Publishing. (1996). *The Knowledge-based economy: Science, technology and industry outlook*. Retrieved from http://www.oecd.org/dataoecd/51/8/1913021.pdf

OECD Publishing, BMWi, & EPO. (2005). *Intellectual property as an economic asset: key issues in valuation and exploitation: Summary Report*. Retrieved from http://www.bmwi.de/BMWi/Redaktion/PDF/C-D/conference-on-intellectual-property-summary,property=pdf,bereich=bmwi,sprache=de,rwb=true.pdf

Orey, M. & Herbst, M. (2006). *Inside Nathan Myhrvold's Mysterious New Idea Machine*. Retrieved from http://www.businessweek.com/magazine/content/06_27/b3991401.htm

Parchomovsky, G., & Wagner, P. (2005). Patent portfolios. *University of Pennsylvania Law Review*, *154*(1), 1–77.

Parr, R. L., & Smith, G. V. (2008). *Intellectual Property: Valuation, Exploitation, and Infringement Damages: Valuation, Exploitation, and Infringement Damages*. Hoboken, NJ: John Wiley & Sons.

Parr, R. L., & Sullivan, P. H. (Eds.). (1996). *Technology licensing: Corporate strategies for maximizing value*. Hoboken, NJ: John Wiley & Sons.

PatentFreedom. (2011a). *Leading Entities by Number of Counterparties and Litigations*. Retrieved from https://www.patentfreedom.com/research-ml.html

PatentFreedom. (2011b). *NPEs with Largest Patent Holdings*. Retrieved from https://www.patentfreedom.com/research-phl.html

Pisano, G. (2006). Profiting from innovation and the intellectual property revolution. *Research Policy*, *35*(8), 1122–1130.

Pitkethly, R. H. (1997). *The valuation of patents: A review of patent valuation methods with consideration of option based methods and the potential for further research* (Judge Institute Working Paper WP 21/97). Retrieved from http://bus6900.alliant.wikispaces.net/file/view/EJWP0599.pdf

Pitkethly, R. H. (2001). Intellectual property strategy in Japanese and UK companies: patent licensing decisions and learning opportunities. *Research Policy, 30*(3), 425–442.

Pluvinage, V. (2011). IP business models - past, present and future. *Intellectual Asset Management Magazine, 48*, 44–52.

Polanyi, M. (1962). *Personal knowledge: Towards a post-critical philosophy* (2. impr., with corr.). London: Routledge & Kegan Paul.

Polanyi, M. (1967). *The tacit dimension*. Garden City, NY: Anchor Books.

Punch, K. F. (2005). *Introduction to Social Research: Quantitative and Qualitative Approaches: Quantitative and Qualitative Approaches* (2. ed.). Thousand Oaks, CA: SAGE Publications.

Rassenfosse, G. de. (in press). How SMEs exploit their intellectual property assets: evidence from survey data. *Small Business Economics*.

Reepmeyer, G. (2006). *Risk-sharing in the pharmaceutical industry: The case of out-licensing*. Heidelberg: Physica-Verlag.

Reepmeyer, G., Gassmann, O., & Rüther, F. (2011). Out-licensing in markets with asymmetric information: The case of the pharmaceutical industry. *International Journal of Innovation Management, 15*(4), 1–41.

Reinhard, M., & Schmalholz, H. (1996). *Technologietransfer in Deutschland: Stand und Reformbedarf*. Berlin: Duncker & Humblot.

Reinhardt, D. (2008). Pre-Licensing Considerations. *Licensing Journal, 28*(5), 15–21.

Reitzig, M. (2003). What determines patent value?: Insights from the semiconductor industry. *Research Policy, 32*(1), 13–26.

Reitzig, M. (2004). Improving patent valuations for management purposes: validating new indicators by analyzing application rationales. *Research Policy, 33*(6-7), 939–957.

Reitzig, M. (2004a). Strategic Management of Intellectual Property. *MIT Sloan Management Review, 45*(3), 35–40.

Reitzig, M. (2004b). The private values of 'thickets' and 'fences': towards an updated picture of the use of patents across industries. *Economics of Innovation & New Technology, 13*(5), 457–476.

Reitzig, M., Henkel, J., & Heath, C. (2007). On sharks, trolls, and their patent prey: Unrealistic damage awards and firms' strategies of "being infringed". *Research Policy, 36*(1), 134–154.

Rings, R. (2000). Patentbewertung - Methoden und Faktoren zur Wertermittlung technischer Schutzrechte. *GRUR, 10*, 839-848.

Rivette, K., & Kline, D. (2000). *Rembrandts in the attic: Unlocking the hidden value of patents*. Boston, MA: Harvard Business School Press.

Roberts, B. (1999). A tale of two patent strategies. *Electronic Business, 10*(25), 79–84.

Rogers, E. M. (2003). *Diffusion of innovations* (5th ed.). New York, NY: Free Press.

Rosenbloom, R. S., & Cusumano, M. A. (1987). Technological Pioneering and Competitive Advantage: The Birth of the VCR Industry. *California Management Review, 29*(4), 51–76.

Rubin, S. (2007). Defending the Patent Troll: Why These Allegedly Nefarious Companies Are Actually Beneficial to Innovation. *Journal of Private Equity, 10*(4), 60–63.

Ryan, P. (2011). *Interview on approval of case study "Acacia Research Corporation"*. Telephone.

Sandburg, B. (2001). *You May Not Have a Choice: Trolling for Dollars*. Retrieved from http://www.phonetel.com/pdfs/LWTrolls.pdf

Sapsed, J., Grantham, A., & DeFillippi, R. (2007). A bridge over troubled waters: Bridging organisations and entrepreneurial opportunities in emerging sectors. *Research Policy, 36*(9), 1314–1334.

Schankerman, M., & Pakes, A. (1986). Estimates of the value of patent rights in European countries during the post-1950 periode. *Economic Journal, 96*(384), 1052–1076.

Scheffer, G. von. (2008). Letter to the editor. *Intellectual Asset Management Magazine, 31*, 5.

Scherer, F. M. (1965). Firm Size, Market Structure, Opportunity, and the Output of Patented Inventions. *American Economic Review, 55*(5), 1097–1125.

Scherer, F. M., & Harhoff, D. (2000). Technology policy for a world of skew-distributed outcomes. *Research Policy, 29*(4-5), 559–566.

Scotchmer, S. (2006). *Innovation and incentives*. Cambridge, MA: MIT Press.

Serafino, D. (2007). *Survey of Patent Pools Demonstrates Variety of Purposes and Management Structures* (KEI Research Note 2007:6). Retrieved from http://keionline.org/content/view/69/1

Serrano, C. J. (2010). The dynamics of the transfer and renewal of patents. *The Rand Journal of Economics, 41*(4), 686–708.

Shapiro, C. (2001). Navigating the Patent Thicket: Cross Licenses, Patent Pools, and Standard Setting. In A. B. Jaffe, J. Lerner, & S. Stern (Eds.), *NBER Book Series Innovation Policy and the Economy. Innovation Policy and the Economy* (1st ed., pp. 119–150). Cambridge, MA: MIT Press.

Sheehan, J., Martinez, C., & Guellec, D. (2004). *Understanding business patenting and licensing: results of a survey*.

Shohert, S., & Prevezer, M. (1996). UK biotechnology: Institutional linkages, technology transfer and the role of intermediaries. *R&D Management, 26*(3), 283–298.

Shrestha, S. K. (2010). Trolls or Market-Makers? An Empirical Analysis of Nonpracticing Entities. *Columbia Law Review, 110*(1), 114–160.

Silverman, B. S. (1999). Technological Resources and the Direction of Corporate Diversification: Toward an Integration of the Resource-Based View and Transaction Cost Economics. *Management Science, 45*(8), 1109–1124.

Smith, M., & Hansen, F. (2002). Managing intellectual property: a strategic point of view. *Journal of Intellectual Capital*, *3*(4), 366–374. doi:10.1108/14691930210448305

Sneed, K. A., & Johnson, D. K. N. (2009). Selling ideas: the determinants of patent value in an auction environment. *R&D Management*, *39*(1), 87–94.

Spulber, D. F. (1999). *Market microstructure: Intermediaries and the theory of firm*. Cambridge: Cambridge University Press.

Stankiewicz, R. (1995). The role of the science and technology infrastructure in the development and diffusion of industrial automation in Sweden. In B. Carlsson (Ed.), *Technological systems and economic performance. The case of factory automation*. Dordrecht: Kluwer Academic Publishers.

Stevens, G. A., & Burley, J. (1997). 3,000 raw ideas = 1 commercial success! *Research Technology Management*, *40*(3), 16–27.

Stigler, G. J. (1951). The Division of Labor is Limited by the Extent of the Market. *Journal of Political Economy*, *59*(3), 185.

Svensson, R. (2007). Commercialization of patents and external financing during the R&D phase. *Research Policy*, *36*(7), 1052–1069.

Tapscott, D., & Williams, A. D. (2008). *Wikinomics: How mass collaboration changes everything* (expan. ed.). London: Atlantic Books.

Taylor, A. (2007). What Does Forum Shopping in the Eastern District of Texas Mean for Patent Reform? *The John Marshall Review of Intellectual Property Law*, *6*(3), 570–589.

Teece, D. J. (1981). The Market for Know-How and the Efficient International Transfer of Technology. *Annals of the American Academy of Political and Social Science*, *458*, 81–96.

Teece, D. J. (1986). Profiting from technological innovation: Implications for integration, collaboration, licensing and public policy. *Research Policy*, *15*(6), 285–305. doi:10.1016/0048-7333(86)90027-2

Teece, D. J. (1998). Capturing Value from Knowledge Assets: The New Economy, Markets for Know-How, and Intangible Assets. *California Management Review*, *40*(3), 55–79.

Teece, D. J. (2000). *Managing intellectual capital: Organizational strategic and policy dimensions*. New York, NY: Oxford University Press.

Tessera Technologies. (2011). *Invensas Acquires ALLVIA 3D-IC Packaging Technology*. Retrieved from http://www.invensas.com/_layouts/newsArticles_manual_print.aspx?ID=619582

The Economist online. (2010). *Patent nonsense: An end to frivolous patents may finally be in sight*. Retrieved from http://www.economist.com/node/15479680/print

Thumm, N. (2004). Strategic Patenting in Biotechnology. *Technology Analysis & Strategic Management*, *16*(4), 529–538.

Tietze, F. (2011). *Managing Technology Market Transactions: Can Auctions Facilitate Innovation?* Cheltenham: Edward Elgar Publishing.

Tietze, F. & Herstatt, C. (2010). *Technology Market Intermediaries and Innovation* (Conference Paper). Imperial College London Business School. Retrieved from http://www2.druid.dk/conferences/viewpaper.php?id=502063&cf=43

Towns, W. R. (2010). U.S. Contingency fees: A level playing field? *WIPO Magazine*, (1), 3–6.

Troy, I. & Werle, R. (2008). *Uncertainty and the Market for Patents*. MPIfG Working Paper 08/2 (MPIfG Working Paper 08/2). Retrieved from http://www.mpi-fg-koeln.mpg.de/pu/workpap/wp08-2.pdf

Tschirky, H., & Escher, J.-P. (2000). Technology marketing: A new core competence of technology-intensive enterprises. *International Journal of Technology Management, 20*(3/4), 459.

van de Vrande, V., Vanhaverbeke, W., & Gassmann, O. (2010). Broadening the scope of open innovation: past research, current state and future directions. *International Journal of Technology Management, 52*(3/4), 221–235.

van Lente, H., Hekkert, M., Smits, R., & van Waveren, B. (2003). Roles of Systemic Intermediaries in Transition Processes. *International Journal of Innovation Management, 7*(3), 247–279.

Vickery, G. (1988). A survey of international technology licensing. *STI Review, 4*, 7–49.

Voss, C., Tsikriktsis, N., & Frohlich, M. (2002). Case research in operations management. *International Journal of Operations & Production Management, 22*(2), 195.

Wang, A. W. (2010). Rise of patent intermediaries. *Berkeley Technology Law Journal, 25*(1), 159–200.

Watson, A. (2010). *Nortel-runners and riders*. Retrieved from http://www.tangible-ip.com/2011/nortel-runners-and-riders.htm

Watson, R. T., Boudreau, M.-C., York, P. T., Greiner, M. E., & Wynn, D., JR. (2008). The Business of Open Soruce. *Communications of the ACM, 51*(4), 41–46.

Wernerfelt, B. (1984). A resource-based view of the firm. *Strategic Management Journal, 5*(2), 171–180.

Westney, D. E. (1993). Cross-Pacific internationalization of R&D by U.S. and Japanese firms. *R&D Management, 23*(2), 171–181. doi:10.1111/j.1467-9310.1993.tb00084.x

Wild, J. (2010a). *America's multi-billion dollar IP marketplace*. Retrieved from http://www.iam-magazine.com/blog/detail.aspx?g=f7cc77e4-9791-42ac-8342-a48086434a59&q=marketplace#search=%22marketplace%22

Wild, J. (2010b). *The "major semiconductor company" that has just sold Acacia a DRAM patent portfolio*. Retrieved from http://www.iam-magazine.com/blog/detail.aspx?g=b97d2e44-c933-42b0-b2f8-eace24817693

Williams, D., & Gardner, S. (2006). Basic framwork for effective responses to patent trolls. *IP Link, 17*(3), 3–5.

Williams, S. (2006). *A Haven for Patent Pirates*. Retrieved from http://www.technologyreview.com/InfoTech-Software/wtr_16280,300,p1.html

Williamson, O. E. (1975). *Markets and hierarchies: analysis and antitrust implications: A study in the economics of internal organization*. New York, NY: Free Press.

Williamson, O. E. (1983). Credible Commitments: Using Hostages to Support Exchange. *American Economic Review, 73*(4), 519–540.

Williamson, O. E. (1985). *The economic institutions of capitalism: Firms, markets, relational contracting*. New York, NY: Free Press.

Winch, G. M., & Courtney, R. (2007). The Organization of Innovation Brokers: An International Review. *Technology Analysis & Strategic Management, 19*(6), 747–763.

Yanagisawa, T. & Guellec, D. (2009). *The emerging patent marketplace* (STI Working Paper 2009/9). Retrieved from http://www.oecd.org/dataoecd/62/55/44335523.pdf

Yin, R. K. (2009). *Case study research: Design and methods* (4th ed.). Thousand Oaks, CA: SAGE Publications.

Yurkerwich, D. (2008). Patent sales and the IP business plan. *Licensing in the boardroom*, 37–40.

Zahra, S. A., & Nielsen, A. P. (2002). Sources of Capabilities, Integration and Technology Commercialization. *Strategic Management Journal, 23*(5), 377–398.

Ziedonis, R. H. (2004). Don 't Fence Me In: Fragmented Markets for Technology and the Patent Acquisition Strategies of Firms. *Management Science, 50*(6), 804–820.

Appendix

Appendix 1: General information of analyzed patent aggregating companies

Patent aggregating company	Company information	Business data	Field of business
Acacia Research Corporation (Acacia)	Headquarters in Newport Beach, CA, USA. Originally incorporated in California in 1993, reincorporated in Delaware in 1999 as venture capital firm. Started patent license business in 2003, IPO in 2003 (NASDAQ: ACTG). Key persons: Paul Ryan (Chairman & CEO), Robert Harris (Director & President) - both founders	Ca. 50 full time employees. Controls 180 patent portfolios containing US patents and certain foreign counterparts. Completed more 1,000 licensing agreements across 99 technology programs	Area of activity: Aggregates mainly US patents from companies in the US, Europe, Asia. Business model: Aggregates patents from high-technology industry that are already in use and enforces these patents.
Alliacense	Headquarters in Cupertino, CA, USA. Founded in 2004. Subsidiary of IP Management service provider TPL Group. Key persons: Mac Leckrone (President), Mike Davis (Senior Vice President, Licensing)	Ca. 30 employees. Controls 224 US patents in 124 patent families	Areas of activity: Aggregates mainly US patents from companies in the US, Europe, Asia. Business model: Aggregates patents from high-technology industry that are already in use and enforces these patents.

Patent aggregating company	Company information	Business data	Field of business
Allied Security Trust	Headquarters in Lambertville, NJ, USA Founded in 2007, among the founding members are Ericsson, Hewlett-Packard, IBM, Intel, Motorola, Oracle, and Philips. Currently 21 members, all operating company with annual revenues of USD 500 million per year and more. Key persons: Dan McCurdy (CEO), Kerry G. Hopkin (CFO), Linda Biel (Vice President)	Eight fulltime employees Has acquired ca. 500 patents	Area of activity: Defensively purchases[1] mainly US patents from companies in the US, Europe, Asia. Members are international companies with headquarters in North America, Asia, and Europe. Business model: Organization without profit orientation, provides members freedom to operate by acquiring patents which may otherwise be asserted against them by a non-practicing entity, members decide which patents are bought, after providing licenses to members patents are resold on the market and funds received are returned to participating members.

Patent aggregating company	Company information	Business data	Field of business
Alpha Patentfonds	Headquarters in Frankfurt, Germany Founded in 2007 by Euram Bank (initiator), Alpha Patentfonds GmbH (portfolio company) and Vevis (sales partner), in 2010 Charles River Associates was mandated for exploitation of patents	Three investment funds (Alpha Patentfonds I - closed Q3/2007, asset under management EUR 32.7 million, number of patent families 164; Alpha Patentfonds II - closed Q4/2008, asset under management EUR 49.3 million, number of patent families 246; Alpha Patentfonds III - Tranche 2008: closed Q4/2008, asset under management EUR 10.3 million, number of patent families 52 and Tranche 2009: closed Q4/2009, asset under management EUR 6.23 million, number of patent families 31) - all funds blind pools and public placement	Area of activity: Aggregates mainly European patents from companies with offices in Germany, Austria, and Switzerland. Business model: Collects funds of investors, aggregates patents from all industries, bundles new patent portfolios, sells or out-license new portfolios.
AlseT IP (AlseT)	Headquarters in New York, NY, USA Founded in 2000, private company Key person: Laurence Rosenberg (Senior Managing Director)	Less than 25 employees	Area of activity: Worldwide. Business model: Acquires patent backed royalty streams from all industries, bundles them to portfolios, and refinances these transactions at the capital market.
Capital Royalty	Headquarters in Houston, TX, USA Founded in 2003, private company Key person: Charles Tate (Chairman, Founding Partner)	Ca. 20 employees Investments range from USD 20 million to USD 200 million (upfront payment to patent owner)	Area of activity: Worldwide. Business model: Primary and secondary market of royalties investments, aggregate royalty streams of patents covering FDA-approved healthcare products, refinances these transactions at the capital market.

Patent aggregating company	Company information	Business data	Field of business
Coller Capital	Headquarters in London, Great Britain Founded in 1990, private equity firm, IP investment group at Coller Capital was set up in 2006 Key person: Peter Holden (Partner, Head IP Investment Group Coller Capital)		Area of activity: Worldwide. Business model: Aggregates patents and exploits the patents in different ways.
CreativE	Headquarters in Europe Founded in a year between 2000 and 2009 as investment vehicle, operations started in 2010	Ca. five employees Three patent portfolios, five to ten licensing agreements in 2010	Area of activity: Mainly Europe, expanding worldwide. Business model: Aggregates patents that cover consumer electronics and are already in use and enforces these patents.
Eco-Patent Commons	Based in Geneva, Switzerland Launched in 2008 by IBM, Nokia, Pitney Bowes, and Sony in partnership with the World Business Council for Sustainable Development (WBCSD) that hosts the Commons	104 patents are donated from twelve companies and universities	Area of activity: Worldwide. Business model: Aggregates patents from donors and provides royalty free licenses to foster research and innovation to protect the environment. No industry focus but patents should provide direct or indirect environmental benefit.

Patent aggregating company	Company information	Business data	Field of business
Fergason Patent Property (Fergason Patent)	Headquarters in Menlo Park, CA, USA Founded in 2001, private company Key person: James L. Fergason (Founder)	Ca. ten employees	Area of activity: Worldwide. Focus on patents covering electronic displays and liquid crystal technology Business model: Aggregates patents that are already in use and enforces these patents, until now without filing lawsuits.
Golden Rice product development partnership (Golden Rice PDP)	Golden Rice Project Management in Freiburg, Germany Scientific details were first published in 2000 Key persons: Peter Beyer (Creator of the technology), Ingo Portrykus (Creator of the technology), Dr Jorge Mayer (Golden Rice Project Manager)	Free licenses of 70 patents are donated by 32 different companies and universities to enable the production of Golden Rice	Area of activity: Developing countries. Business model: Non-profit organization, aggregates patents and free licenses to enable the production of Golden Rice and makes the technology available to resource-poor farmers in developing countries.
IgniteIP	Headquarters in Mountain View, CA, USA Founded in 2002, private company Key persons: Brandon Williams (Managing Director), Vlad Dabija (Managing Director)	Ca. eight employees Investments range from USD 500,000 to USD 2 million.	Area of activity: Mainly United States. Business model: Aggregates patents from software, cleantech, and biopharma from early-stage prospective technologies, advances patents through contract R&D and exploits technology by selling or licensing.

Patent aggregating company	Company information	Business data	Field of business
Intellectual Ventures	Headquarters in Bellevue, WA, USA Founded in 2000, private company Key persons: Founders are Nathan Myhrvold (CEO), Edward Jung (CTO), Peter N. Detkin (Vice Chairman), Greg Gorder (Vice Chairman)	Ca. 800 employees Has aggregated a patent portfolio of more than 35,000 US and international patents and patent applications Generated more than USD 2 billion in licensing revenue from 30 major licensees	Area of activity: Worldwide. Business model: Aggregates patents from various industries, either already in use or embryonic technologies, and applies different exploitation strategies.
IP Holdings	Headquarters in Suffern, NY, USA Founded in 2000 as idea incubator of General Patent Corporation and spun off in the same year. Still affiliated with General Patent Corporation. Managers and employees work for both companies. Key person: Alexander Poltorak (Chairman and CEO of General Patent Corporation)	IP Holdings has aggregated patents and bundled them to seven portfolio companies	Area of activity: Aggregates mainly US patents from companies in the US, Europe, Asia. Business model: Aggregates patents from life science and electrical engineering, invests in development and incubation, and assists in the commercialization of novel and promising technology. Additionally, IP Holdings provides IP-related financial services and IP brokerage.

Patent aggregating company	Company information	Business data	Field of business
IP Navigation Group (IP Navigation)	Headquarters in Dallas, TX, US Founded in 2005, privately held, main company of a conglomerate that covers the entire IP value chain (identification of patents through Patent Calls, Consulting of patent owners through IP Navigation, enforcement of patents through several companies as e.g., Gemini IP, Plutus IP, Orion IP, Taurus IP, etc.) Key person: Erich Spangenberg (CEO and Founder)	Conglomerate is operated by eight employees IP Navigation Group has aggregated 41 patents from single inventors and research institutions and generated 543 licensing agreements.	Areas of activity: Aggregates mainly US patents from companies in the US, Europe, Asia. Business model: Aggregates patents from various industries that are already in use and enforces these patents.
MPEG Licensing Administration (MPEG LA)	Headquarters in Greenwood Village, CO, USA Founded in 1996 Key persons: Lawrence A. Horn (President and CEO), JP Gascon (CFO), Alexis DeVane (General Counsel)	Ca. 15 employees Operates licensing programs consisting of more than 5,000 patents with ca. 130 licensors and 5,000 licensees. Generates revenues of around USD 1 billion per year	Area of activity: Worldwide. Business model: Developed a many-to-many licensing model where multiple users are able to acquire essential patent rights from multiple patent holders in a single transaction as an alternative to negotiating separate licenses. Performs this approach in electrical engineering and life science industry.

Patent aggregating company	*Company information*	*Business data*	*Field of business*
Open Invention Network (OIN)	Headquarters in Durham, NC, USA Founded in 2005, founding members are IBM, NEC, Novell, redhat, Philips, and Sony. These companies still finance the aggregation activities. Key Person: Keith Bergelt (CEO)	Ca. three employees 210 aggregated patents, all covering the system software Linux	Area of activity: Mainly United States. Business model: Non-profit organization, aggregates (acquiring or receiving donations) patents and offers free licenses to users who further develop the open source software Linux. The users do not pay royalties but commit not to use patents for blocking Linux.
Techquity Capital (Techquity)	Headquarters in Austin, TX, USA Founded in 2008, private company Key person: Abha Divine (Founder and Managing Director)	Ca. five employees	Area of activity: Mainly United States. Business model: Aggregates portfolio of high quality patents covering embryonic and prospective technologies, advances technology and develops further, and licenses them broadly into the market.
Papst Licensing GmbH & Co.KG (Papst Licensing)	Headquarters in St. Georgen, Germany Based on manufacturing company Papst-Motoren GmbH & Co. KG, founded in 1992 as Papst Licensing GmbH to monetize patent rights. Since about 2000 in patent aggregating business. Key persons: Georg Papst (CEO), Daniel Papst (CPO), Constantin Papst (CFO), Tobias Kessler (senior counsel)	About 14 employees Patent portfolio contains about 140 patents, 20% of these patents are acquired, with over 150 license agreements, mainly with companies in IT, electrical engineering, and electronics.	Area of activity: Worldwide. Business model: Aggregates patents from various industries that are already in use and monetizes/ enforces these patents.

Patent aggregating company	Company information	Business data	Field of business
Patent Invest Fond	Headquarters in Pullach, Germany. Founded in 2005 by Finance System (initiator), Patenthandel Portfolio I (portfolio company) and Credit suisse (sales partner), partner in the selection and exploitation process are Steinbeis TIB and PATEV	Fund was closed in Q4/2005, asset under management ca. EUR 20 million, number of patents 30-128, minimum invested capital EUR 50,000, blind pool and private placement	Area of activity: Worldwide. Business model: Collects funds of investors, aggregates patents from all industries, bundles new patent portfolios, sells or out-licenses new portfolios.
Patent Select	Headquarters in Schönefeld, Germany. Founded in 2006 by Clou Partners and Deutsche Bank (initiators), IP Bewertungs AG (patent management and selection company)	Patent Select is the umbrella term for 3 investment funds that follow the same model (Patent Select I - closed Q4/2006, asset under management EUR 24.5 million, numbers of patents ca. 12, asset pool, public placement; Patent Select II - closed Q3/2007, asset under management EUR 32.7 million, number of patents ca. 12, asset pool, public placement; Patent Portfolio I- closed Q4/2007, asset under management EUR 130 million, number of patents ca. 22, partly blind pool, public placement)	Area of activity: Worldwide. Business model: Collects funds of investors, aggregates patents from all industries, enhances and develops technologies and patents, sells or out-licenses advanced technologies or patent portfolios.
Pete Invest MedTech (Pete Invest MedT)	Headquarters in the US. Founded in 1999 as investment platform of equity capital firm Pete Invest	Ca. 15 employees working for Pete Invest MedTech (more than 100 working for Pete Invest) 40 investments in pharmaceutical royalties streams, invested capital USD 1.3 billion	Area of activity: Worldwide. Business model: Acquires patent backed royalty streams from pharmaceutical products, bundles them to portfolios, and refinances these transactions at the capital market.

Patent aggregating company	Company information	Business data	Field of business
Rembrandt IP Management (Rembrandt IP)	Headquarters in Bala Cynwyd, PA, USA Founded in 2004, private company Key persons: Paul Schneck (Chairman), Michael Johnson (President), John Garland (Vice President Rembrandt Solutions)	Ca. ten employees Has aggregated 235 US patents in 174 patent families	Area of activity: United States. Business model: Aggregates patents from various industries that are already in use and enforces these patents.
Royalty Pharma	Headquarters in New York, NY, USA Founded in 1996 with first deal in 2000, private company Key persons: Pablo Legorreta (Founder and CEO), Susannah Gray (Executive Vice President and CFO)	14 employees Owns royalty interests in 17 approved and marketed biopharmaceutical products, five products in clinical trials and/or under review with the FDA In 2010 Royalty Pharma owned royalty revenues of USD 808.5 million	Area of activity: United States. Business model: Acquires patent backed royalty streams from biopharmaceutical products, bundles them to portfolios, and refinances these transactions in securitization transactions at the capital market.
RPX Corp. (RPX)	Headquarters in San Francisco, CA, USA Founded in 2008, IPO in May 2011 (NASDAQ: RPXC)$ Key persons: John Amster (CEO), Geoffrey Barker (Chief Operating Officer), Eran Zur (President)	Ca. ten employees RPX operates with a membership structure. Amongst its ca. 80 clients are Dell, Google, IBM, Microsoft, Nokia but also smaller firms with venture backed status. Annual fees depend on the annual revenues. Members do not pay extra for patent acquisition activities. Has aggregated more than 1,600 US and international patents and invested more than USD 260 million.	Area of activity: United States. Business model: For-profit organization, provides members freedom to operate by acquiring patents already in use, members cannot decide which patents are bought, after providing licenses to all members patents are resold on the market. Patents cover consumer electronics, software, media, communications, and semiconductors.

Patent aggregating company	Company information	Business data	Field of business
Sipro Lab Telecom (Sipro Lab)	Headquarters in Montreal, Canada Founded in 1994, private company Key person: Nathalie Beaudoin (Licensing Director)	Ca. 16 employees Administers five patent portfolios covering mobile wireless technologies, with more than 200 licensees	Area of activity: Worldwide. Business model: Administers licensing program where multiple users are able to acquire essential patent rights from multiple patent holders in a single transaction as an alternative to negotiating separate licenses.
Via Licensing Corporation (Via Licensing)	Headquarters in San Francisco, CA; USA Founded in 2003 and is a wholly owned subsidiary of Dolby Laboratories, Inc. Key persons: Jean-Michel Bourdon (President), Nate Alvord (Vice President, Licensing and Program Management),	Ca. 35 employees Manages eleven patent portfolios with ca. 100 patent owners and 800 licensees	Area of activity: Worldwide. Business model: Administers licensing program where multiple users are able to acquire essential patent rights from multiple patent holders in a single transaction as an alternative to negotiating separate licenses.

Source: interviews, annual reports, company documents, articles, internet documents.

[1] According to Biel (29.11.2011) Allied Security Trust does not consider itself a patent aggregator. They do not hold patents and as a result will never asset them themselves.

Appendix 2: Interview - Guideline

Setting of patent aggregating companies

1. What is your position and what is your job description?
2. Please quantify: year of founding, location of headquarters, number of employees, number of acquired patents/ patent applications, capital invested in patent acquisitions, average price per aggregated patents.
3. How did your business involve and what is the history of your company?
4. How do you finance the patent aggregating activities? Have you raised a fund? Who invests in your company or the fund? How do you find investors? What are the funds characteristics?

Strategy of patent aggregating companies

5. How would you describes the company's business model?
6. How would you describe the company's overall strategy?
7. How would you describe your strategy compared to your competitors? How do you differentiate from your competitors?
8. Why do you aggregate patents and what do you do with the patents you have aggregated?
9. How would you describe your unique selling proposition?
10. How could your business model develop within the next five years?

Organization of patent aggregating companies

11. How would you describe your internal competencies? In which area and for which tasks do you have internal resources?
12. Which services do you offer your clients?
13. Do you work with external partners? How does the cooperation work? How would you describe your business model?

Process of patent aggregating companies

14. What is your industrial focus and from which industries do you acquire patents?
15. Please describe the typical legal and technological characteristics a patent you aggregate have.
16. Do you aggregate sole legal rights or do you also transfer technology and knowledge?

17. How do you find potential patents? Who are the original patent owner you aggregate patents from? How do you detect them? How do you approach them?
18. What are the main motivations for patent owners to utilize you?
19. How do you compensate the original patent owner?
20. Which other problems do you solve for the patent owner?
21. Who is the owner of the patent after you have aggregated the patents? Do original patent owners have remaining rights on their patents?
22. How do you evaluate the patents? Which methods do you use to value them? What are important evaluation criteria?
23. How would you describe your value adding activities?
24. What are your exploitation strategies for the patents you have aggregated? What type of exploitation do you conduct?

Appendix 3: List of interviews

Company	Interview partner	Position	Place	Date
Acacia Research	Paul Ryan	CEO	Phone Interview	Mar 9, 2010
			Phone Interview	Jun 22, 2010
Allied Security Trust	Ms. A.	Confidential	Phone Interview	Mar 19, 2010
			Phone Interview	Dec 6, 2010
	Dan McCurdy	CEO	Phone Interview	Jun 10, 2011
Alpha Gasser Patentverwertungs AG	Christian Frey	Head of Patent Commercialization	Phone Interview	Jun 15, 2010
			St. Gallen	Aug 11, 2009
Alpha Patentfonds Management	Bernd Herrmann	CEO	Interview	Jun 27, 2010
Capital Royalty	David Carter	Principal	Phone Interview	Jul 29, 2010
	Mike Weinman	Managing Director	Phone Interview	Apr 21, 2011
Coller Capital	Peter Holden	Head of IP Investment Group	Gothenburg	Sep 08, 2009
CreativE	Mr. K	Management	Phone Interview	Confidential
Deutsche Bank	Frank Rohwedder	Global Banking Asset Finance and Leasing	St. Gallen	Jul 29, 2009
			Phone Interview	Jun 16, 2010
			Phone Interview	Feb 28, 2011
Fergason Licensing	Charles McLaughlin	Managing Partner	Phone Interview	Jun 15, 2010
Finance System	Andreas Fritsch	CEO	Phone Interview	Apr 7, 2010
General Patent Corporation	Alec Schibanoff	Vice President Marketing	Phone Interview	April 07, 2011
IgniteIP	Brandon Williams	Managing Director	Phone Interview	Jul 16, 2010
Intellectual Ventures	Mr. F.	Confidential	Phone Interview	Confidential
	Mr. F.	Confidential	Phone Interview	Confidential
IP Bewertungs AG	Guido von Scheffer	Director	Hamburg	Aug 8, 2009
	Stephan Lipfert	Director IP Management	Interview	Jun 25, 2010

Appendix

Company	Interview partner	Position	Place	Date
IP Navigation Group	Erich Spangenberg	Founder & CEO	Phone Interview	Nov 04, 2010
			Phone Interview	Mar 31, 2011
IP Navigation Group Europe	Lucia Alvarado	Vice President	Phone Interview	Apr 04, 2011
MPEG LA	Bill Geary	Vice President	Phone Interview	Apr 13, 2011
Open Invention Network	Keith Bergelt	CEO	Phone Interview	Feb 22, 2010
	Mr. B	Management	Phone Interview	Feb 17, 2011
Papst Licensing GmbH & Co KG	Daniel Papst	Managing Director, Co-Founder & CPO	Phone Interview	Apr 01, 2011
			Phone Interview	Feb 25, 2011
			Phone Interview	Dec 1. 2010
Pete Invest MedTech	Mr. W	Partner	Phone Interview	Aug 11, 2010
			Phone Interview	Mar 21, 2011
Patent Freedom	Chris Reohr	CEO	Phone Interview	Mar 3, 2010
PATEV Associates	Michael Beyer,	COO	Phone Interview	Jun 14, 2010
			Phone Interview	May 3, 2011
RPX Corp.	Kevin Barhydt	Vice president	Phone Interview	Nov 23, 2010
	Thomas Westerlund	Vice President, Structured Acquisitions	Phone Interview	Mar 23, 2011
Steinbeis TIB	Steffen Schnitzer	Senior Patent Manager	Zurich	May 25, 2010
			Phone Interview	June 22, 2010
Steinbeis TIB	Bernd Singer	Head of Key Account	Zurich	May 25, 2010
			Phone Interview	Mar 22, 2010
Techquity Capital	Abha Divine	Managing Director	Phone Interview	Jul 22, 2010

MIX
Papier aus verantwortungsvollen Quellen
Paper from responsible sources
FSC® C105338

If you have any concerns about our products,
you can contact us on
ProductSafety@springernature.com

In case Publisher is established outside the EU,
the EU authorized representative is:
**Springer Nature Customer Service Center GmbH
Europaplatz 3, 69115 Heidelberg, Germany**

Printed by Libri Plureos GmbH
in Hamburg, Germany